37.50

# DISCRIMINATION AND PUBLIC POLICY
## IN NORTHERN IRELAND

Learning Resources
Centre

# DISCRIMINATION AND PUBLIC POLICY IN NORTHERN IRELAND

Edited by

ROBERT J. CORMACK
AND
ROBERT D. OSBORNE

CLARENDON PRESS · OXFORD
1991

Oxford University Press, Walton Street, Oxford OX2 6DP

Oxford New York Toronto
Delhi Bombay Calcutta Madras Karachi
Petaling Jaya Singapore Hong Kong Tokyo
Nairobi Dar es Salaam Cape Town
Melbourne Auckland

and associated companies in
Berlin Ibadan

Oxford is a trade mark of Oxford University Press

Published in the United States
by Oxford University Press, New York

© the several contributors 1991

British Library Cataloguing in Publication Data
Data available

Library of Congress Cataloging in Publication Data
Discrimination and public policy in Northern Ireland/edited by
Robert J. Cormack and Robert D. Osborne.
p.   cm.
Includes bibliographical references (p.    ) and index.
1. Discrimination in employment—Northern Ireland.   2. Affirmative action programs—
Northern Ireland.   I. Cormack, R. J. (Robert J.)
II. Osborne, R. D. (Robert D.)
HD4903.5.G7D57   1991   331.13'3'09416—dc20   91-11492
ISBN 0-19-827519-6

Typeset by Rowland Phototypesetting Ltd,
Bury St Edmunds, Suffolk
Printed and bound in
Great Britain by Bookcraft (Bath) Ltd,
Midsomer Norton, Avon

15.6.94

*Dedicated to our friend and colleague*

PROFESSOR JOHN H. WHYTE, 1928–1990

*and to*

CHRISTINA M. CORMACK, 1906–1990

# *Acknowledgements*

We have been working in the area of equal opportunities in Northern Ireland for most of our professional careers; it is, therefore, perhaps invidious to pick out certain colleagues for particular mention in the conduct of our work. But, that said, there are colleagues who stand out in the support they have afforded us and/or the rigour with which they have conducted, both formal and informal, debates with us. We have dedicated the book to John Whyte who, more than anyone we know, attempted to understand all sides of the issues in Northern Ireland, and did it in the most scrupulous and assiduous manner. Thereafter, we have greatly benefited and appreciated times spent with Rosalie Abella, Robert Cooper, John Darby, John Edwards, Anthony Gallagher, Jeremy Harbison, and the late Roy Wallis.

Henry Hardy, of Oxford University Press, has our gratitude for his forbearance. Pam Bleekes typed and retyped all the chapters of this book. Our constant demands and revisions were always 'no problem'.

# Contents

# *Figures*

# Tables

# *Abbreviations*

CHS      Continuous Household Survey (the Northern Ireland equivalent of the General Household Survey)

CRE      Commission for Racial Equality

DED      Department of Economic Development (Northern Ireland)

EOC      Equal Opportunities Commission (Northern Ireland)

EOU      Equal Opportunities Unit (Northern Ireland Civil Service)

FEA      Fair Employment Agency (Northern Ireland)

FEC      Fair Employment Commission (Northern Ireland)

NICS      Northern Ireland Civil Service

PSI      Policy Studies Institute

SACHR      Standing Advisory Commission on Human Rights (Northern Ireland)

YOP      Youth Opportunities Programme

YTP      Youth Training Programme (the Northern Ireland equivalent of the Youth Training Scheme)

# Contributors

GEORGE APPLEBEY, Lecturer in Law, University of Birmingham.

GERALD CHAMBERS, Senior Research Officer, Research and Policy Planning Unit, Law Society.

ROBERT COOPER, Chairman, Fair Employment Commission.

ROBERT CORMACK, Senior Lecturer and Head of the Department of Sociology and Social Policy, The Queen's University of Belfast.

URSULA DUFFY, Senior Assistant Statistician, Policy Planning Research Unit, Department of Finance and Personnel, Northern Ireland.

EVELYN ELLIS, Senior Lecturer in Law, University of Birmingham.

Professor DAVID EVERSLEY, formerly Professor of Population and Regional Studies, University of Sussex; and Chief Strategic Planner, Greater London Council.

Dr ANTHONY GALLAGHER, Research Officer, Centre for the Study of Conflict, University of Ulster.

Dr JAMES GILLAN, Deputy Principal Statistician, Policy Planning Research Unit, Department of Finance and Personnel, Northern Ireland.

Dr JEREMY HARBISON, Under-Secretary, Department of Health and Social Services, Northern Ireland.

WILLIAM HODGES, Permanent Secretary, Department of Agriculture, Northern Ireland.

Dr CHRISTOPHER McCRUDDEN, Fellow, Lincoln College, University of Oxford.

Dr LIZ McWHIRTER, Head of Social Statistics and Research Branch, Department of Health and Social Services, Northern Ireland.

Dr ROBERT OSBORNE, Director of the Centre for Policy Research and Senior Lecturer in Social Policy, University of Ulster.

DENISE THOMPSON, Senior Assistant Statistician, Department of Economic Development, Northern Ireland.

# Introduction

*Robert J. Cormack and Robert D. Osborne*

Religious sectarianism has underpinned life in Northern Ireland for a considerable time. Like racism and sexism it helps create barriers to the achievement of equality of opportunity. Protestants and Catholics, blacks and whites, males and females have, to use Max Weber's term, differential 'life chances'. For two hundred years advanced liberal democracies have been evolving policies to enhance and promote equality of opportunity. From Adam Smith's advocacy of the role of the free market in opening up opportunities in the eighteenth century to the post-World War II political consensus, derived from John Maynard Keynes and William Beveridge's state interventionist policies, the liberal goal of equal opportunity has been pursued in a variety of ways. In earlier years passive forms were favoured, whereas recently more active forms of intervention have been encouraged (Cormack and Osborne, 1983). The reversion to free-market-based methods of delivering equality of opportunity in the programme of 'Thatcherism', which seeks to replace the state interventionism of 'Butskellism', has not been fully implemented in Northern Ireland, as this book will demonstrate (Gaffikin and Morrissey, 1990).

More than twenty years of what are colloquially called the 'troubles' in Northern Ireland have provided an important test case for the efficacy of policies to deal with divisions in society. The 'national question' (i.e. whether or not Northern Ireland should remain part of the UK or be incorporated into the Republic of Ireland) has, for many years, dominated the political agenda. But, for most people building their lives in the Province, the degree to which their life chances are enhanced or impeded by their community of origin is the everyday reality. In particular, policies dealing with creating and sustaining equality of *employment* opportunities are central.

The increasing salience and visibility of ethnic divisions throughout the world draws Northern Ireland more to the centre of studies which hope to gain policy insights through the

comparative understanding of societies and structures (a point
we make in Chapter 1). The new 1989 fair employment legislation in Northern Ireland, dealing with discrimination on the
grounds of religion, is in advance of similar policies dealing
with sex and race discrimination in Europe, and compares
favourably with policies in North America.

In Chapter 1 we explore the development of fair employment
from the early 1970s through to the implementation of the
recent legislation in the 1990s. In this task we make frequent
reference to the influence of both Canadian and US legislation.
While the goal of equal opportunities is a political universal in
liberal democracies, the circumstances in which it is to be
achieved vary considerably. Glib comparisons are less than
useful, and potentially dangerous. Fruitful comparisons require detailed background understanding of the characteristics
of the problem in particular jurisdictions.

The chapters which follow attempt to provide the detailed
understanding of the problem of discrimination on the basis of
religion as it is confronted in Northern Ireland. In Chapter 2 we
outline the profiles of the two communities in terms of employment and unemployment using data from the population censuses, recent surveys, and investigations undertaken by the
Fair Employment Agency. David Eversley, in Chapter 3,
assesses the demographic factors at work. In the process, he
demolishes a number of arguments which tend towards the
type which 'blames the victim'. For example, blaming Catholics for having large families, which the labour market cannot
support with jobs, is nonsense since the decision to have
children was made some considerable time ago when economic
conditions were much better.

Education in Northern Ireland is almost completely segregated on the basis of religion, with a Catholic system and a state
(*de facto* Protestant) system. The labour market consequences
for children emerging from these two systems have, in recent
years, become a major focus of research. In Chapter 4, with
Anthony Gallagher, we summarize the important findings in
this area. The implications of post-school training programmes
for the two communities are explored in Chapter 5 by Liz
McWhirter and colleagues.

Thereafter, chapters focus on fair employment policies *per se.* In Chapter 6, Gerald Chambers provides evidence from a major survey of employers in the Province. If one single conclusion is to be drawn from this survey it is that a voluntary approach, inviting employers to participate freely in the delivery of equal employment opportunities, rarely, if ever, works. Even the Northern Ireland Civil Service (NICS) did not institute proper procedures until it was subjected to a Fair Employment Agency investigation. However, the subsequent development of the Equal Opportunities Unit within the NICS can, justifiably, be held up as a model for other employers to follow. Jeremy Harbison and William Hodges outline the evolution of this unit and its work in Chapter 7.

The responsibility for ensuring that employers adhere to fair employment policies and practices has been the responsibility of the Fair Employment Agency. Robert Cooper was chairman of the Agency from its inception in 1976. He has now taken over as the chairman of the Fair Employment Commission con- stituted under the new legislation in 1990. In Chapter 8, Cooper looks back over the successes and failures of the FEA. In Chapter 9, the FEA is assessed, from a legal standpoint, by George Applebey and Evelyn Ellis, both in terms of its work and in terms of the legislation under which it operated.

Christopher McCrudden, in Chapter 10, documents the passage of the new fair employment legislation through Parlia- ment. In so doing, a number of potential weaknesses in the drafting of this legislation are identified. Chapters 1, 8, 9, and 10 provide discussions of the lessons of the past together with an assessment of what the future of the new legislation may offer.

Rarely does legislation command a broad political con- sensus; that much is clear from Chapter 10. Nevertheless, 'warts and all', the new Northern Ireland fair employment legislation is undoubtedly in the forefront, internationally, of policies in this vitally important area. Political legitimacy depends greatly on a sense of fairness in the operation of the labour market. As Eastern Europe evolves new political struc- tures, the ethnic diversity of many of these newly emerging democracies is likely to ensure that the liberal concern with equality of employment opportunities will emerge as central, just as it is in Western Europe and notably in North America. The ethnic conflict already visible in the new Europe, together with the amplification of such conflicts in North America, suggests that Northern Ireland should be considered a major comparator in studies of ethnic conflict and the policies initiated to tackle such problems.

# 1

# Disadvantage and Discrimination in Northern Ireland

*Robert J. Cormack and Robert D. Osborne*

## ETHNIC IDENTITY AND EMPLOYMENT DIFFERENTIALS

The conflict in Northern Ireland is usually described as a conflict between Protestants and Catholics. These labels have the unfortunate effect of conveying a sense of Northern Ireland as the scene of an atavistic example of a holy war. However, given the recent and continuing changes in Eastern Europe, it is possible and credible to suggest that Northern Ireland may shortly find itself heading a long list of European and other societies riven by ethnic conflicts. To Cyprus and the Lebanon, to Belgium and Canada, to the problems of the Basques and the Bretons, can now be added the ethnic conflicts in Yugoslavia, Romania, and Bulgaria, not to mention the varied and mounting ethnic conflicts within the Soviet Union. The comparative study of such societies may come to reveal Northern Ireland as an exemplar of the social and political problems encountered. Over the twenty or more years of the recent 'troubles' in Northern Ireland various policies have been tried and tested to improve community relations and to tackle some of the root causes of the conflict. As Northern Ireland moves from being considered a 'deviant' to a 'typical' case of a type of problem widespread throughout the new Europe, such policies will become of ever-greater interest to social scientists studying the contemporary world.

In the study of such conflicts the focus of this book is narrow but fundamental: the distribution of employment between the two Northern Irish communities, and the policies enacted to try to ensure fairness in this distribution. Equality of employment opportunities is a fundamental issue in all societies, but no more so than in societies where ethnic groups compete for power and influence. This chapter seeks to provide an overview of the evolution of policies designed to deliver equality of employment opportunities.

In Northern Ireland, as in the emerging conflicts in a number of other European societies, ethnicity and religiosity are closely intertwined. Jenkins (1988: 310), following Weber, states: 'ethnic identity and the boundaries of ethnic groups are situationally defined by the actors concerned, subject to negotiation and redefinition within the constraints of history and present circumstances'. In homes, in schools, in workplaces, in the mundane interactions of everyday life, ethnic identities are never very far from the surface (Harris, 1972; Burton, 1978). On the other hand, individuals are often concerned to project their individuality and uniqueness and, in the process, reject being pigeon-holed by religion. In the 1971 population census 9% refused to state a religion; in 1981 this non-response rate had risen to 19% (Morris *et al.*, 1985). The new fair employment legislation, introduced in 1989, requires employers with more than 25 employees to monitor the religious composition of their workforces and to submit this information, annually to the Fair Employment Commission. Anticipating problems from asking the direct question—'what is your religion?'—the legislation has suggested indirect methods of assigning people to a religion, most notably on the basis of primary school attended. The latter approach points up contradictory approaches taken by government, in that the Education Reform Order of 1989 gives considerable encouragement to the expansion of integrated schools. The new fair employment legislation seems to separate and divide people on the basis of religion in order to enhance greater structural equality, whereas a key objective of the educational reforms is the provision of integrated education with the goal of promoting better community relations.

In many ways this apparent policy contradiction reflects an ambivalence in popular discussion of the issue. Many will argue that one's religion is a private matter and of no concern to an employer. Taken a little further, it has been argued that if an employer does not know an applicant's religion then he or she cannot be guilty of discrimination. On the other hand, a more pragmatic view, embraced by government in the framing of the new fair employment legislation, is, that all kinds of clues and cues may be used to determine the religion of a job applicant. At the structural level, since the employment profiles of the two communities reveal substantial differences between Protestants and Catholics, the delivery of equality of opportunity requires the monitoring of workforces on the basis of religion.

In Northern Ireland, a person's community of origin greatly affects who that person will marry, the area they will live in, the schools their children will attend, often the companies they will approach when seeking employment, and the cemeteries they will be buried in. Ethnic identity is built upon all the mundane interactions of everyday life. In the new Europe it is less and less difficult to put the case for the power of ethnic identity to mobilize people. The cultural and symbolic resources underpinning ethnic identities are highly emotive: flags, anthems, language, calendar dates (often reflecting the 'invention of tradition' (Hobsbawm and Ranger, 1983)). In the face of such powerful resources the mobilization of people on the basis of class, which involves a more cerebral understanding of social structures and patterns of exploitation, faces a difficult task. Miliband (1987: 344), commenting on class analysis, offers the following thoughts on Northern Ireland:

It is tempting to see this struggle as a 'sectarian' one, based purely upon religious and ethnic grounds. But it is not particularly 'class reductionist' or an exaggerated form of 'economic determinism' to suggest that a basic cause of the antagonism is the attempt by Protestant workers to safeguard their already precarious and even dire material situation from what they take to be a major threat from an even more deprived minority, with both sides distinguished by religion, tradition, culture, historic memories and mutual grievances. Undoubtedly, this economically-generated antagonism is rationalized and expressed in terms which are far removed from their economic roots. These terms soon acquire solidity and substance and therewith autonomy, and thus become powerful ideological constructs in their own right. People subscribe passionately to these constructs and come to define much of their 'social being' in the constructs' terms . . . The cord which attaches economic position to ideological construct is a very long one and runs through very rugged terrain. It is often buried deep, and it may snap altogether.

The history of attempts to build a non-sectarian socialism in Northern Ireland bears witness, not so much to the difficulties of identifying the economic roots of the problem, but rather to the immense difficulty of building and sustaining working-class politics across the sectarian divide (Bew *et al.*, 1979).

The topic of ethnicity cannot be left without comment on the degree to which religion *per se* underpins the identity of Protestants and Catholics. This has been the subject of much debate in recent years. There are those who see the labels 'Protestant' and 'Catholic' as merely 'flags of convenience' (e.g. Bell, 1976). As Aughey (1989: 3) describes this position: 'Religion is not the cause of the divisions and violence in Northern Ireland. It is merely a manifestation of divisions which are held to be real and material not ideal and spiritual.' On the other hand, there are those who take religion to be the fundamental component in the identities of the two communities. Bruce (1986: 249, 258), in his study of Ian Paisley, argues bluntly that the 'Northern Ireland conflict is a religious conflict' and that for Protestants 'beyond evangelical Protestantism, no secure identity is available'. Catholics are in a somewhat different position, he argues, in that they have an unambiguous national identity available to them which has become detached from a religious base. Aughey (1989: 7) challenges this analysis by suggesting that 'Bruce overestimates the political secularism of nationalism and undervalues its religious significance. Similarly he underestimates the secularism of unionism and severely overplays its religious particularism.' Aughey argues that the Republic of Ireland is 'a Catholic state' and hence Catholic nationalism is not usefully portrayed, as in Bruce's work, as a secular ideology. Protestant Unionism is merely 'caricatured' by tying it down, in a reductionist manner, to evangelical Protestantism; for Aughey, secular strands within Unionism have been and are important.

This debate, and other recent work in the area (Wright, 1973, 1987; Bell, 1976; Nelson, 1984; Todd, 1987; Fulton, 1988), useful and interesting as it is, seems at one remove from a level at which we might understand how the patterns of employment advantage and disadvantage arose and continue to be maintained. How do people find jobs? How do employers recruit workers? In Northern Ireland, as elsewhere, the common practice is for employers to recruit 'by word of mouth' using informal networks. These networks may be based on a variety of activities: church, family and neighbourhood contacts, leisure activities (golf clubs, the Masons, Knights of Columbanus, Boys Brigade, Gaelic Athletic Association, schools' past pupil associations, etc.). Analyses of the 1911 population census suggest that there were substantial differences within the Protestant community between Presbyterians, Methodists, and members of the Church of Ireland, with the latter group's profile more similar to that of Catholics than of Presbyterians (Cormack and Rooney, n.d.). On the face of it, church-going probably provided people with the key location through which they established social relationships which, in turn, determined the scope and extent of employment opportunities. Then as now, employers want a cheap and reliable means of recruiting staff. They want to know that they will have a good and reliable worker. Doing a fellow church member a favour in giving a job to his son or daughter carries with it the obligation on the father to try to ensure that his son or daughter is punctual, conscientious, and hard-working. Formal recruiting methods using advertisements, short-listing, interviewing, and the gathering of references do not provide the employer with the same level of assurance that derives from the mutual obligations arising through the use of informal networks. But even as formal methods have become more common there is good reason to suspect that informal mechanisms continue to operate under the surface (Maguire, 1990).

Whether being Protestant or Catholic is merely an ethnic label or a confessional identity matters little in this context. What is important is the issue of the advantage or disadvantage associated with being *identified* as a member of one or other of the two communities.

The employment profiles of the two communities demonstrate a long-standing pattern of Catholic disadvantage. The period of industrialization and urbanization in the nineteenth century laid the basis for contemporary employment patterns (Hepburn, 1983). In the recent past these differences have been one of the sources of agitation in the 1960s (Campaign for

Social Justice, 1969; Cameron Commission, 1969) and one of the key issues which the politics of Direct Rule in the 1970s and 1980s has had to address, as will be discussed below. The chapters in this book continue the work we first developed in *Religion, Education and Employment: Aspects of Equal Opportunity in Northern Ireland* (Cormack and Osborne, 1983). Here we explore the extent of continuing Catholic employment disadvantage and various aspects of policy interventions. While gender and class divisions are as pertinent in Northern Ireland as elsewhere they will only be considered here inasmuch as they pertain to the divisions deriving from religion. In policy terms, the approach taken to religious divisions in Northern Ireland is informing contemporary debate in Britain on both sex and race discrimination and the mechanisms and agencies required to tackle these problems (CRE, 1990).

## EVOLUTION OF FAIR EMPLOYMENT IN NORTHERN IRELAND

Discrimination in Northern Ireland has long been alleged in the areas of the franchise, the location of electrical boundaries, the allocation of public housing, and employment. Whyte (1983) has produced the most authoritative attempt to review the extent of discrimination covering the period of devolved administration from 1921 to 1968. His review confirms the existence of discrimination but demonstrates that malpractices were particularly prevalent in certain geographical areas and in certain activities. He is especially critical of Unionist-controlled local authorities in the west of Northern Ireland. Whyte concludes his review by offering a list of 'demerit' where he considers the evidence of discrimination was strongest and most widespread: electoral practices, public employment, policing, private employment, public housing, and regional policy. In sum: 'The unionist government must bear its share of responsibility. It put through the original gerrymander which underpinned so many of the subsequent malpractices, and then despite repeated protests did nothing to stop those mal-practices continuing' (Whyte, 1983: 31).

Whyte evaluated the period of the devolved Unionist-controlled government. In the period immediately prior to the assumption of Direct Rule in 1972, the Unionist government, partially as a response to its own hesitant reform agenda, but more directly under pressure from Westminster—which was increasingly worried by civil disturbances—instituted changes in local government, the local government franchise, and the administration of housing, health, and personal social services. The reform package also included the creation of the offices of Parliamentary Commissioner for Administration (PCA) and the Commissioner for Complaints (CC). Alleged mal-administration, including religious discrimination, in govern-ment was under the purview of the PCA, and in other public bodies by the CC (Birrell and Murie, 1980). After the institu-tion of Direct Rule a working party, subsequently chaired by William van Straubenzee MP, was created to 'consider what steps, whether in regard to law or practice, should be taken to counter religious discrimination where it may exist in the private sector of employment in Northern Ireland' (MHSS, 1973: 1). The group, which consisted mainly of employers and trade union representatives, reported in 1973. It recorded a 'general acceptance of the validity of the basic assumption underlying our terms of reference that religious discrimination exists to some degree as a fact of life in employment in Northern Ireland today' (MHSS, 1973: 2). It went on to reject quotas on moral grounds (they would involve discrimination), on a prac-tical basis (problems of defining appropriate catchments), but above all because 'the effect of quotas would not be to reconcile the communities in Northern Ireland but on the contrary [would] reinforce and in some measure . . . perpetuate the divisions between them' (MHSS, 1973: 12).

Northern Ireland
Key Developments in Northern Ireland
Fair Employment Policy

| | |
|---|---|
| van Straubenzee Working Party Report | 1973 |
| Fair Employment Act | 1976 |
| Fair Employment Agency established | 1976 |
| Guide to Manpower Policy and Practice | 1978 |

As can be seen above the van Straubenzee Working Party's
report set in train government concern for fair employment in
the Province beginning with the 1976 Fair Employment Act,
which endorsed the essence of the Working Party's proposals.
Under the legislation the Fair Employment Agency (FEA),
was created with two main duties: the elimination of discrimination on the grounds of religious belief or political opinion,
and the promotion of equality of opportunity between those of
different beliefs. To carry out these duties the FEA was given
three main powers and responsibilities:

1. The investigation of individual complaints of discrimination.
2. The power, under Section 12 of the Act, to investigate
   employment practices and to judge whether or not they
   accord with principles of equality of opportunity.
3. To conduct research germane to fair employment in
   Northern Ireland and to pursue an educational programme to inform employers of what are considered to be
   fair and proper employment practices.

In conducting Section 12 investigations the FEA was enjoined
to follow the *Guide to Manpower Policy and Practice*, prepared
under the Act, by the then Department of Manpower Services.
The *Guide* provided recommendations to employers for establishing policies and practices which, if adopted, would
promote equality of opportunity (DMS, 1978). Under Section
12 investigations the FEA could direct firms to implement an

'affirmative action' programme in the event of a finding of 'failure to afford equality of opportunity'. In some instances this has involved the use of goals and timetables. The *Guide* was substantially revamped, particularly in relation to advice on monitoring employees, in 1987 (DED, 1987*a*). In addition, tenders for government contracts were restricted to firms registered with, and holding a certificate from, the FEA.

The FEA, from its inception, faced daunting circumstances. Unionist politicians opposed the initial legislation and were suspicious of its work thereafter, perceiving the objectives of the legislation as hostile to their community's interests. During the 1970s and early 1980s trade unions and Nationalist politicians rarely referred to the issue of discrimination in the context of fair employment legislation. Nationalist perceptions of the FEA tended to be of the Agency as either a toothless sop to provide respectability to Direct Rule, or more supportively as an underfunded body tackling a major problem without adequate government support. Direct Rule administrations, particularly under Labour in the 1970s, concentrated on job creation, achieving 'parity' in the provision of services in Northern Ireland with those in Britain, and on security; in the process, equality of opportunity was given a low priority. As a result not only was the FEA modestly funded and staffed but, more damagingly, the government itself did little to demonstrate equality of opportunity in the Northern Ireland Civil Service (NICS). Set alongside these problems, the continuing violence and the virtual collapse of manufacturing employment in the early 1980s provided a context for the FEA's work which could hardly have been less propitious (Osborne and Cormack, 1989*a*).

Nevertheless, how effective was the FEA? Applebey and Ellis, in Chapter 9, offer a comprehensive evaluation of the work of the Agency. McCrudden (1983) has criticized the professionalism and organization of the FEA. Other commentators have suggested that the number of investigations undertaken prompts favourable comparison with the CRE and EOC in Britain (see Applebey and Ellis, Chapter 9). The Standing Advisory Commission on Human Rights (SACHR) reviewed the performance of the Agency with regard to the conducting of

14      *Cormack and Osborne*

Section 12 investigations and the issuing of remedial instructions to employers and, while rehearsing the criticisms and
rebuttals in detail, did not make an evaluation (SACHR,
1987). Many of the critics have, however, failed to distinguish
between the limitations imposed by the legislation and the
effectiveness of the FEA.

## THE PRESSURES FOR REFORM

### *Northern Ireland Civil Service*

A number of factors led to the decision to reform. The first
relates to developments within Northern Ireland. The FEA's
investigation of the Northern Ireland Civil Service (NICS),
completed in 1983, stimulated the government to institute an
equal opportunity monitoring system for the NICS (FEA,
1983). This system, which monitors the NICS in terms of
religion, gender, and disability, represents the most sophisticated monitoring system in the UK. It now covers recruitment,
promotions, etc. and has reported regularly since 1986 (see
Harbison and Hodges, Chapter 7; Osborne, 1990). The creation of this system represented the government putting its own
house in order. Furthermore it resulted in key civil servants
becoming sensitized to the issues and to the failure of the public
sector as a whole to respond positively to the legislation. At the
same time, continuing FEA investigations were revealing the
inequalities in employment profiles between Protestants and
Catholics in numerous organizations and the failure of employers, public and private, to take voluntary action to implement equality of opportunity. Evidence from research studies,
some sponsored by the FEA, confirmed the continuity of
employment disadvantage for Catholics (Cormack and Osborne, 1983). The publication in 1985 of the results of the
government's own new Continuous Household Survey (CHS)
confirmed the higher experience of unemployment for Catholics and was accompanied by a statement from the then
Secretary of State for Northern Ireland, Douglas Hurd, announcing the institution of a review, to be undertaken by
officials, into how the existing approach to equality of oppor-

tunity could be made 'more comprehensive, consistent and effective' (NIO, 1985). It is clear, then, that the administration in Northern Ireland was by the mid-1980s responding, albeit belatedly, in concrete ways to the legislation in order to effect major change.

### The MacBride Principles and Irish Americans

The role of Irish-Americans has been crucial in translating this growing awareness into legislative proposals from a Conservative government notoriously reluctant to advance equal opportunity policy in the UK as a whole. Pressure from Irish-American groups has taken two forms. The first was the attempt by the Irish National Caucus, in 1983, to prevent the US government from buying aircraft from the Belfast aircraft company Shorts on the grounds that the firm was accused of actively discriminating against Catholics. The British government was forced to intervene with the US government to counteract this allegation. As a result, Shorts, under the guidance of the FEA, instituted a major change in their employment practices, which has reportedly begun to improve the representation of Catholics in the workforce (*Belfast Telegraph*, 1988). The more direct pressure from Irish-Americans has, however, been in relation to the MacBride Principles.

The origins of the MacBride Principles lie in an article in the *Brooklyn Spectator* in 1983. This article suggested that, following the Sullivan Principles applied to firms operating in South Africa, a similar set of principles should be applied to US firms operating in Northern Ireland. Harrison Goldin, Comptroller of the City of New York at the time, assigned a member of his staff, Patrick Doherty, to investigate discrimination in Northern Ireland and to come up with a set of Sullivan-type principles for Northern Ireland. Doherty, in collaboration with activists in Irish affairs in the United States and in Northern Ireland, produced a set of principles to which Sean MacBride was invited to give his name. The principles were issued in late 1984 signed by Sean MacBride, Dr John Robb (a Northern Irish surgeon), Inez McCormack (a Northern Irish trade union leader), and Father Brian Brady (a West Belfast priest).

### THE MACBRIDE PRINCIPLES

1. Increasing the representation of individuals from under-represented religious groups in the workforce, including managerial, supervisory, administrative, clerical and technical jobs.

2. Adequate security for the protection of minority employees both at the workplace and while travelling to and from work.

3. The banning of provocative religious or political emblems from the workplace.

4. All job openings should be publicly advertised and special recruitment efforts should be made to attract applicants from under-represented religious groups.

5. Layoff, recall and termination procedures should not, in practice, favour particular religious groups.

6. The abolition of job reservations, apprenticeship restrictions and differential employment criteria, which discriminate on the basis of religion or ethnic origin.

7. The development of training programmes that will prepare substantial numbers of current minority employees for skilled jobs, including the expansion of existing programmes and the creation of new programmes to train, upgrade and improve the skills of minority employees.

8. The establishment of procedures to assess, identify and actively recruit minority employees with potential for further advancement.

9. The appointment of a senior management staff member to oversee the company's affirmative action efforts and the setting up of timetables to carry our affirmative action principles. (Rubenstein, 1986: 18–19)

The Principles have been mobilized in two ways to involve US companies operating in Northern Ireland: shareholder resolutions, and state and city legislation. The shareholder campaign began in 1985. Proposals, in various forms but deriving from the MacBride Principles, were submitted for consideration at annual meetings. Companies were asked either to adhere to the MacBride Principles or to conduct in-depth reviews of their operations in Northern Ireland and to report back to shareholders. The other plank in the campaign began in New York City, where two of Goldin's funds, the Employees' Retirement System and the Teachers' Retirement System, adopted resolutions in 1985 directing the comptroller to encourage companies in the retirement systems' portfolios to adopt and implement the MacBride Principles.

Massachusetts, in 1985, was the first state to enact legislation, followed by other Eastern seaboard states down to Florida in 1988, together with some mid-West states beginning with Minnesota in 1988. On balance, legislation has been aimed at getting companies to monitor and, where necessary, improve their employment practices rather than at legislation which could lead to divestment.

The Principles are in themselves relatively innocuous. In some instances the FEA insisted on stronger measures being taken by firms subject to investigation under Section 12 than is required by the principles. However, legal advice taken by the FEA suggested that principles 1, 7, and 8 could imply 'reverse discrimination' and hence be illegal under the 1976 Act.

In 1986, MacBride issued 'amplifications' to the principles, as follows.

Principle 1: 'A workforce that is severely unbalanced may indicate *prima facie* that full equality of opportunity is not being afforded all segments of the community in Northern Ireland. Each signatory to the MacBride Principles must make every reasonable lawful effort to increase the representation of underrepresented religious groups at all levels of its operations in Northern Ireland.'
Principle 7: 'This does not imply that such programmes should not be open to all members of the workforce equally.'
Principle 8: 'This section does not imply that such procedures should not apply to all employees equally.' (Rubenstein, 1986: 18–19)

The British government has fought the MacBride Principles all the way. In one sense this may seem strange. The Ford Motor Company, responding to the MacBride initiative, conducted a study, chaired by the Vice-President for Employee and External Affairs, of their plant in Northern Ireland. This study team reported:

In conducting its review, the Ford study team tried to consider the wider purpose of those who accept the MacBride Principles. Dr MacBride's 1986 amplifications of the principles make it apparent that he did not intend to encourage unlawful actions in Northern Ireland. Yet legal uncertainties remain as to whether some of the MacBride Principles can be implemented lawfully. The study team decided to deal with this problem by refining a statement of fair employment policy for Northern Ireland which it believes adopts the substance of the MacBride Principles while avoiding legal and operational obstacles to implementation. (Ford Motor Company, 1987: p. ii)

Why could not the British government have welcomed the MacBride Principles as worthy fair employment guidelines and suggested, in an approach similar to that of Ford, incorporating the essence of them in what was at that time the draft of a Consultative Document on Fair Employment (DED, 1986)?

Legal advice to the FEA on the potential illegality of some of the Principles, under the 1976 Fair Employment Act, was certainly a consideration. But of much greater moment were three further factors: the linking and simplistic equating of the circumstances in South Africa and Northern Ireland; worries about disinvestment; and the nature of the groups advocating the passage of the MacBride Principles through state and city legislatures, in particular, Sean MacBride, Father McManus, and the Irish National Caucus. Booth and Bertsch (1989), in their review of the MacBride Principles for the Washington-based Investor Responsibility Research Center, state:

MacBride won the 1974 Nobel Peace prize for work at the United Nations. He was a co-founder of the human rights group Amnesty International and served as its chairman from 1961 to 1974. From 1948 to 1951, he was the Republic of Ireland's Minister of External Affairs, and before that, in the early 1930s, he was chief of staff of the IRA . . . MacBride's name has worked both for and against the principles. On the one hand, it has given them the stature of a serious initiative and bolstered them against the charge that they are just an attempt by Comptroller Goldin to gain electoral support from Irish-Americans. It has also disposed some Irish-American groups to support the principles because of their admiration for MacBride. On the other hand, MacBride's early personal history has prompted others to regard the principles as another republican nationalist tool to embarrass the British or even to destabilize the Northern Irish economy and bring about the collapse of British direct rule. (Booth and Bertsch, 1989: 56)

By opposing the Principles in the United States the British government put itself in an impossible position. In media terms it was an 'own goal'. To many Americans the Principles seem reasonable and generally in line, if somewhat weaker, than affirmative action policies in the US. The opposition of the British government to such Principles suggested to many a lack of commitment in resolving Catholic employment disadvantage in Northern Ireland. However, every political party in Northern Ireland, except Sinn Fein, followed the government's lead. Notably, two prominent Catholic politicians, John Hume, leader of the SDLP, and Paddy Devlin, formerly in the SDLP and now a trade union leader, have argued forcefully against the Principles both in Northern Ireland and in the US. The gist of this opposition lies in the need to create jobs and the implications of disinvestment by US firms operating in Northern Ireland. Inward investment, it is argued, is only likely to be hindered by the naïve linking of Northern Ireland to South Africa.

The MacBride campaign undoubtedly kept the pressure on the British government throughout the period during which the new legislation was formulated and enacted. Has the campaign now served its manifest purpose? The Investor Responsibility Research Center concluded in December 1989: 'Even if one accepts the arguments made by the British in 1986, it would appear that the claimed legal conflicts between MacBride and the law are narrowed or eliminated by the changes made by the new fair employment law' (IRRC, 1989: 21). It should be remembered here that the MacBride principles only apply to US firms operating in Northern Ireland, of which twenty-seven had been identified and targeted for shareholder action by the end of 1989 (IRRC, 1989). The new legislation imposes duties on all Northern Irish employers, such as monitoring the religious composition of the workforce, and carries penalties not canvassed by MacBride, in particular loss of government grants and contracts for failure to comply with the new legislation. However, the MacBride campaign continues. More clearly than ever, the campaign must now be evaluated on the degree to which it hinders or enhances investment and job-creation in Northern Ireland.

Equally, the new FEC must ensure fair employment. But it must also be recognized that it is not a job-creation agency and new employment is what is urgently required in Northern

Ireland, in particular to help redress past patterns of Catholic employment disadvantage. The scale of the problem is suggested by the fact that in the six-year period from June 1983 to June 1989 the number of persons in employment increased by 18,600 while the number of unemployed fell by only 500. It has been calculated that it will take the creation of three additional jobs in order to reduce the unemployed figures by just one (Gudgin and Roper, 1990). Factors such as the continuing high birth-rate, which results in greater numbers entering the labour market, the increasing participation of women in the labour market starting from a comparatively low base, and a pattern of migration where fewer people choose to leave Northern Ireland and where some jobs in Northern Ireland may be taken by the return of previous out-migrants, contribute to making a difficult problem even more severe. In addition, public-sector employment in Northern Ireland forms a much higher proportion of total employment than elsewhere in the UK, with a correspondingly smaller private sector—a private sector which continues to show few signs of the rate of growth experienced in other regions (NIEC, 1990). Given the importance of the public sector, the mismatch between areas of Catholic population concentration and the location of job opportunities suggests that the state must continue to carry a major responsibility for the generation and location of employment both through the influence it exerts on the private sector and directly through public-sector jobs.

### The Standing Advisory Commission on Human Rights' Fair Employment Review

Continuing pressure for reform came from the Standing Advisory Commission on Human Rights (SACHR). The SACHR was established by Parliament under Section 20 of the Northern Ireland Constitution Act of 1973 with the purpose of 'advising the Secretary of State on the adequacy and effectiveness of the law for the time being in force in preventing discrimination on the grounds of religious belief or political opinion and in providing redress for persons aggrieved by discrimination on either ground' (SACHR, 1987: 1).

However, it took the SACHR twelve years to tackle the issue of discrimination in employment (Osborne, 1981). In January 1985 it announced its decision to undertake a major review 'to examine whether in practice the law gives adequate protection against discrimination on the grounds of religious belief or political opinion and whether there exists equality of opportunity in Northern Ireland' (SACHR, 1987: p. iii). This announcement followed after the publication of the government's Consultative Paper on fair employment. To aid its deliberations the SACHR commissioned a number of studies by the Policy Studies Institute (PSI) in London, which were eventually published by the PSI (Smith, 1987*a*, 1987*b*; Chambers, 1987). The PSI reports underpinned the SACHR's conclusions in this area. *Employment and Unemployment* (Smith, 1987*a*) confirmed many of the basic patterns of advantage and disadvantage of Protestants and Catholics in the labour market revealed in earlier research (Cormack and Osborne, 1983; Osborne and Cormack, 1986, 1987). The SACHR then drew selectively on the PSI work for its own report on fair employment (SACHR, 1987), published late in 1987 some months before the government's White Paper on fair employment.

The SACHR evaluated eleven factors, identified from previous research, as having a possible contributory effect on the employment and unemployment profiles of the two communities. Here we will note each factor in turn and the SACHR's conclusion on the contribution of this factor to the differentials. However, various chapters in this book explore these factors in considerably greater depth than we attempt here.

1. Regional factors. The argument here is that Catholics tend to be concentrated in areas with fewer job opportunities, particularly in the west of the Province, and that this locational factor accounts for much of the unemployment differential. The SACHR concluded that this factor made a 'clear but small contribution to the overall differential' (SACHR, 1987: 26).

2. Differential labour markets. The suggestion is that Protestants and Catholics often enter separate labour markets, to the disadvantage of Catholics. While both communities, especially in areas of high residential segregation, tend to have generated their own communal services, from retail shops to

the professions, such segregated employment patterns should balance out with each community 'looking after its own'. Evidence from the 1971 population census and subsequent surveys suggests patterns of over and under-representation of the two communities in certain industries, e.g. Protestants overrepresented in energy and engineering industries, Catholics overrepresented in construction. However, the SACHR concluded that this factor 'cannot . . . properly be regarded as an explanation of the established differentials as opposed to a means of expressing one way in which they are manifested' (SACHR, 1987: 27).

3. Class differences. Class differences between Protestants and Catholics, with Catholics more likely to be found in the lower social classes, may contribute to higher levels of Catholic unemployment since unemployment is higher in these lower social orders. But, as with the segregated labour market argument, the SACHR again concluded that class differences were more an expression of the overall differentials than a cause of them.

4. Differences in education. Catholics gain fewer formal educational qualifications and tend to be biased towards arts and humanities subjects rather than to science and technology; these differences then have an effect on their opportunities in the labour market. Insufficient evidence was available to fully quantify this effect but the SACHR concluded that 'some degree of inequality in employment opportunities in certain sectors may reasonably be attributed to this factor' (SACHR, 1987: 30). The SACHR subsequently commissioned further research in this area.

5. Differences in attitudes to work. Here the argument is that Catholics have a lower commitment to work than Protestants. SACHR found 'the work ethic is equally strong in both the Protestant and the Catholic sections of the community' (SACHR, 1987: 30).

6. Age. The differences in the age distribution in the two communities, with Catholics more highly represented in the younger age groups, it has been argued, contribute to higher levels of Catholic unemployment. This argument depends for its validity on there being an uneven unemployment rate

throughout the age groups to the disadvantage of the younger groups. This, of course, is the case. However, if full equality of opportunity were to pertain the unemployment differential would slowly decline; if the economy became more buoyant, especially for young school leavers, the differential would much more rapidly decline.

7. Family size. The tendency of Catholics to have larger families, it is argued, contributes to higher levels of Catholic unemployment. Heads of large families tend to experience unemployment more than heads of smaller families. However, SACHR concluded that the reasons for this finding are unclear. SACHR was not convinced that social security benefits, particularly for those with large families, necessarily acted as a disincentive to seeking employment.

8. The chill factor. Here the argument is that Protestants and Catholics tend not to want to seek employment in workplaces considered hostile or located in hostile areas. The PSI (Smith, 1987*b*) survey evidence showed that 'fear of intimidation or hostility in an area perceived to be the territory of the other side [was] a potent factor' (SACHR, 1987: 33; see Chapter 6).

9. Security occupations. Catholics have excluded themselves from seeking employment in security occupations, most notably with the security forces. The SACHR suggests that a form of the chill factor may operate here in that Catholics may feel they will be ostracized by their own community or suffer reprisals from paramilitaries if they joined the police or prison service. Also, the SACHR suggested, it may be 'due in part to a reluctance among many Catholics to give full support to the security services in a State which they feel has not given them full recognition or permitted them to participate fully in the processes of government' (SACHR, 1987: 34).

10. The black economy. The argument here is that unemployment has been exaggerated and that, in fact, many 'unemployed' have either occasional or permanent work. The SACHR found no way of assessing the extent of the black economy and hence drew no conclusion on the contribution of this factor to the overall differentials between the two communities.

11. Religion. Finally, the SACHR attempted to assess the degree to which the differentials were the product of direct or indirect discrimination based on religion. Religion was treated as a residual factor introduced when all the other factors have been taken into account. The SACHR decided:

It cannot be concluded that this residual element is or is not due to unlawful direct or indirect discrimination. In the absence of other plausible explanations, however, it is reasonable to suppose that religion is the most likely explanatory factor which accounts for at least some if not most of the residual element. (SACHR, 1987: 35)

Smith, the author of the PSI study *Employment and Unemployment* (1987a), had carried this residual argument much further. In his view, all differentials which cannot be accounted for by other factors which he was able to quantify, must be the result of religious discrimination. We criticized Smith on this approach and a minor, rather acrimonious, debate ensued (Cormack and Osborne, 1989). The only part of this debate worth rehearsing here is Smith's lack of rigour in his use of the term 'discrimination'. In our view, for statistical and conceptual reasons, it is not valid to claim discrimination as the only remaining factor. For example, the author of the most recent PSI study of race relations in Britain states:

The 1966 PEP study revealed that racial discrimination existed on a substantial scale, but the programme of research centred on the 1974 survey demonstrated that there were other factors that resulted in the unfair treatment of black people in Britain, factors that were not just the result of intentional acts of racial discrimination: the policies and practices of employers and other organizations were found often to work against the interests of black people, even though those policies and practices had developed when there were few black people in Britain. (Brown, 1985)

The author of the 1974 study was David Smith (1977). It seems odd that Smith's careful and considered qualifications concerning racial discrimination in Britain did not constrain him from asserting that discrimination is the root cause of the problem in Northern Ireland. 'Discrimination' is an emotive term. In Northern Ireland, as in Britain, it makes sense to use it in a precise fashion. Our practice, over the years (Cormack and

Osborne, 1983), has been to try to distinguish between direct discrimination, where individual maltreatment on the basis of religious affiliation takes place, and indirect discrimination, where disadvantage results from actions or practices which of themselves are not necessarily designed to be biased, an approach indistinguishable from that applied by Smith in his work on Britain's black population. Direct and indirect discrimination often amplify an already existing pattern of disadvantage. For example, in inner cities, racial/ethnic groups suffer social and economic disadvantages; inner-city whites in Birmingham and Protestants on the Shankill suffer; but Birmingham inner-city blacks and Catholics on the Falls suffer an additional disadvantage brought on them by both direct and indirect discrimination. Just as Smith recognized such a phenomenon in Britain, so too should he have been prepared to find it in Northern Ireland.

Our view of Smith's work, it would seem, is shared by Sir Charles Carter, the president of PSI at the time of the Northern Ireland studies. Wilson reports:

it is quite improper, as Charles Carter has observed, to use the term 'discrimination' as though it meant 'any residual difference observed between two communities when selected plausible causes of that difference have been eliminated'. Not only may there be some quantifiable causes that have been overlooked but there are others that cannot be directly quantified. Unfortunately the unwarranted assertion made on the basis of the PSI investigation may lead to greater suspicion and ill-will in a community which already has a more than adequate quota of both. (Wilson, 1989: 114)

The SACHR's policy recommendations went beyond the proposals contained in the Consultative Paper but were substantially incorporated in the Fair Employment Act. Broadly, the government and the SACHR agreed on asserting the importance of selection and promotion on the basis of merit. It is evident that underlying this approach there was an attempt to make clear that 'positive discrimination', or as the SACHR termed it, 'reverse discrimination', was not considered appropriate to circumstances in Northern Ireland for the time being. The SACHR defined reverse discrimination as follows: 'Reverse discrimination occurs where, in order to redress an existing imbalance in the religious composition of a workforce, a person is preferentially selected or promoted over at least one better qualified individual of a different religion' (SACHR, 1987: 70). The SACHR considered reverse discrimination as 'politically unacceptable' and 'fundamentally unjust' but left the way open for a future consideration of its possible appropriateness when 'other affirmative action measures have been tried and found to have failed' (SACHR, 1987: 70–1).

However, the SACHR defined the 'merit principle' in a much less restrictive manner than the government. The SACHR endorsed extensive affirmative action measures allowing, for example, employers to take outreach measures designed to enhance the representation of underrepresented sections of the community at both the application and appointment stages. Latterly, in the Fair Employment Act, as we shall see below, the government went much of the way to endorse this position but stopped short at the point of advocating religion-specific targeting of outreach measures.

### The Anglo-Irish Agreement

The final factor behind the reform proposals has been the pressure brought to bear by the Irish government through the Anglo-Irish Agreement. Although discrimination only became a substantial issue on the agenda of the Inter-Governmental Conference during 1987, since then the Irish government has taken a close interest in the evolving legislation; this is despite the fact that the Irish government's own record on equal opportunities policies is less than impressive.

### TOWARDS REFORM: THE NEW LEGISLATION

As these processes and pressures forced the British government to consider reform, Northern Ireland officials began to search for relevant models for a new policy. In this search there was little to be gained from the traditional reaction of policy-makers in Northern Ireland looking east 'to the mainland'. National policy on race and sex discrimination remained broadly the same as the original fair employment legislation and there were no proposals for reform. Officials therefore looked west to North America.

Affirmative action policies in the United States have been the subject of concerted criticism over the years and the focus of an ultimately unsuccessful attempt, at least during the Reagan administration, to dismantle key aspects of provision (Glazer, 1988). Much of the criticism has been directed towards the concept of 'positive discrimination' and the use of 'quotas'. Although quotas have been employed only to tackle the most egregious cases of discrimination, the contentiousness of US policies has centred on the 'unfairness' of the 'reverse discrimination' implied in the use of quotas (BNA, 1986). Canadian policy statements in the area, however, firmly rejected 'positive discrimination' and 'quotas' in favour of 'positive action' and 'positive remedies'. Hence, Northern Irish civil servants found the evolving Canadian approach potentially more congenial and less contentious, particularly to the Conservative government of the time. As reforms would have to have Cabinet approval, the Canadian policy offered a model that was likely to be more politically acceptable than that to be found in the United States.

### The Influence of Canadian Legislation

The political actions of the Canadian government in the 1960s to address the political alienation of 'francophones', particularly in the Province of Quebec, had resulted in the Official Languages Act of 1969 and the promotion of a bilingual nation. An equal opportunities policy for 'francophones' in the Canadian federal civil service had by 1986 produced a representation of 'francophones' in line with their representation in the wider population. The success of these policies both encouraged and frustrated advocacy groups representing women, the disabled, native persons, and 'visible minorities'. The Abella Commission (1984) reviewed the employment circumstances of four groups—women, native persons, the disabled, and visible minorities—and assessed available policy options. Abella found little support for positive discrimination involving quotas and also argued that the term 'affirmative action' was subject to widespread confusion. She urged the adoption of the term 'employment equity'. Most of Abella's recommendations were endorsed in the 1986 Employment Equity Act. Under this legislation, federally regulated employers and Crown Corporations with 100 or more employees are required to undertake annual monitoring of their workforce for the four designated groups—women, native persons, disabled persons, and visible minorities. The monitoring must include occupational and salary details and all recruitment and wastage. These annual returns, after checking by the government department responsible for monitoring, are passed to the Canadian Human Rights Commission (CHRS). Employers are also required to prepare an annual employment equity plan specifying goals and timetables for implementation. The CHRC can use these data under the Act to institute investigations under procedures established by the 1977 Canadian Human Rights Act (CHRC, 1988*a*). Moreover, the judgement of the Canadian Supreme Court in the *Action Travail des Femmes* v. *Canadian National Railways* case of 1987 means that CHRC can institute goals and timetables as part of its directions to particular employers (CHRC, 1988*b*). (Some commentators viewed the Court's judgement as imposing 'temporary quotas' while others viewed it as ordering 'hiring goals'.) Alongside federal employment equity legislation, a federal contractors programme, applying to those employing more than 100 persons and bidding for contracts of $200,000 or more, was instituted in 1986. Contractors must establish internal goals and timetables for achieving employment equity and are subject to compliance reviews, and ultimately may be excluded from bidding for government contracts. (See Cormack and Osborne, 1990, for further analysis of Canadian material.)

Senior Northern Ireland civil servants kept in close touch with these Canadian developments in the period from 1984 to 1986 and there is no doubt that Canadian policy informed the emergence of the Northern Ireland reform proposals. However, as the drafting of the legislation progressed, the Canadian influence gave way to issues peculiar to Northern Ireland.

## THE NORTHERN IRELAND REFORM PROPOSALS

The development of new fair employment proposals for Northern Ireland began with the publication in 1986 of a Consultative Paper outlining options and possibilities (DED, 1986),
followed by a White Paper (HMSO, 1988*a*) in 1988 indicating
more precisely the government's proposals, the Fair Employment Bill (Northern Ireland) (HMSO, 1988*b*) tabled in the
House of Commons in December 1988, and finally the Fair
Employment (Northern Ireland) Act (HMSO, 1989*b*) which
came into force in January 1990.

### The Consultative Paper

The Consultative Paper set out the basic principles for reform:

1. the rejection of quotas;
2. the centrality of the merit principle;
3. the unacceptability of improving the circumstances of one
   community at the expense of the other.

It summarized in detail the research on employment differentials between Protestants and Catholics but was far less
detailed on gender and the disabled. A number of options,
together with possible legislative changes, were canvassed:

1. changing the 'Declaration of Principle and Intent' to
   one of 'Practice' and accepting tenders for government
   contracts only from those so certified;
2. providing initial financial assistance for the private sector
   to introduce monitoring and better practices in relation to
   religious monitoring (and possibly for gender and disability as well) and taking powers to introduce grant
   denial in relation to religious underrepresentation;
3. placing a statutory duty on the public sector to practise
   equality of opportunity in employment on the basis of
   the procedures set out in the proposed Declaration of
   Practice;
4. establishing an advisory unit in government to give guidance on equality of opportunity in employment in relation
   to religious equality (and possibly gender and disability
   as well); and

5. restructuring the institutional arrangements by establishing either a new Fair Employment Commission which
would concentrate solely on religion (the 'single-
dimension' option), or a new Equal Employment Opportunities Commission to deal with religion, gender, and
disability (the 'multi-dimensional' option).

In advance of new legislation it was proposed to set up a fair
employment support scheme to assist private sector employers
introducing new practices outlined in the new 'Guide to Effective Practice' (DED, 1987*a*). This scheme was introduced in
March 1988 and provides free consultancy advice to private-
sector employers on implementing recommendations contained in the Fair Employment Code of Practice (issued by
DED under the new Act). By July 1990, 900 applications for
assistance had been received and government had spent almost
£¾ million on the scheme.

The Consultative Paper, while outlining these measures, did
not propose the strengthening of the powers of enforcement of
the existing agencies. To a great extent the government proposed to continue to rely on the voluntary compliance of
employers. Considerable debate followed the publication of
these proposals. In particular, the failure to extend mandatory
monitoring to the private sector was criticized. Women's
groups and the Equal Opportunities Commission (EOC) in
Northern Ireland did not favour the multi-dimensional approach, judging that an inevitable concentration on religion
would overwhelm women's interests. The multi-dimensional
approach, which anticipated the possible eventual harmoniza-
tion of the law in relation to religion, gender, and disability,
shows the most obvious influence of evolving Canadian policy;
indeed it is doubtful whether such a proposal would have been
made without some officials clearly being impressed by both
the Abella proposals and the Employment Equity Act.

## The SACHR Review

In the period after the Consultative Paper the most significant
development was the review of religious anti-discrimination
provisions undertaken by the SACHR (SACHR, 1987). The

SACHR report has been described as 'the most impressive analysis of the policy issues which underlie discrimination law which has ever been published in the United Kingdom' (EOR, 1988: 24). As we have noted above, the White Paper (HMSO, 1988*a*) responded to some of the criticisms of the Consultative Paper and endorsed many of the SACHR's recommendations.

## The White Paper

The White Paper concentrated on religious discrimination and rejected the previously canvassed multi-dimensional approach. It proposed the introduction of a legal requirement to monitor, the explicit prohibition of indirect discrimination, the referral of individual discrimination cases to industrial tribunals, the use of 'outreach' goals and timetables in affirmative action remedies, and a form of contract compliance. The White Paper was subjected to critical assessment by McCrudden (1988*b*), a member of the SACHR, who highlighted a number of issues which, in his view, the White Paper dealt with inadequately. These included:

1. employers should be protected if they voluntarily engaged in affirmative action measures;
2. indirect discrimination should be more carefully defined;
3. the legislation should encourage employers to adopt goals and timetables; and
4. contract compliance proposals should be comprehensive.

## The Fair Employment (NI) Bill

The Fair Employment (Northern Ireland) Bill was published in December 1988. To a great extent it carried forward both the strengths and weaknesses of the White Paper. Perhaps the most notable feature of the passage of the Bill through both the Commons and the Lords was the extent to which the government indicated its willingness to consider points and amendments raised by the Opposition. The promises of amendments led to both the Labour Party and the Social, Democratic and Labour Party voting in favour of the Bill at the end of the Third

Reading. This rather unusual openness can be readily understood when one considers the overriding political need for the
government to be able to proclaim the new legislation as
representing an all-party consensus on the best policy response
to this aspect of the Northern Ireland problem. Such a consensus would greatly enhance the government's challenge to
the MacBride Principles in the US. However, the promised
amendments did not materialize in the Lords in forms which
met the approval of the Labour Party, and Labour withdrew
their approval of the Bill when it finally returned to the
Commons. McCrudden, a former SACHR member and an
adviser to Kevin McNamara (Labour's shadow Secretary of
State for Northern Ireland), documents in detail in Chapter 10
the issues which Labour found sufficiently unsatisfactory to
vote against the final passage of the Bill.

### THE FAIR EMPLOYMENT (NI) ACT

The Fair Employment (Northern Ireland) Act 1989 substan-
tially strengthens the provisions made under the 1976 Act. The
Department of Economic Development (DED) published *Key
Details of the Act* in August 1989 with a foreword by Mrs
Thatcher. This glossy publication was widely distributed
throughout Northern Ireland and internationally. Since this is
what the government considers to be the major features of the
Act it is worth quoting from it at length.

The new Fair Employment (NI) Act 1989 shows that the Government is determined to eradicate job discrimination and to ensure the
active practice of equality of employment opportunity in Northern
Ireland . . . It is the most radical fair employment law ever passed by
the United Kingdom Parliament . . . In summary, it puts new duties
on employers to ensure the active practice of fair employment.
Employers must register, monitor their workforces and regularly
review their recruitment, training and promotion practices. They
must take affirmative action measures and set goals and timetables
where necessary. There are both criminal fines and economic
sanctions—involving loss of business and grants—for those guilty of
bad practice. There are two new enforcement bodies, the Fair Employment Commission (which replaces the present Agency and in-
herits its investigatory powers) and the Fair Employment Tribunal.

In addition, indirect discrimination, whether intentional or not, is outlawed. (DED, 1989: 5)

All private-sector employers with more than 25 employees must register with the FEC; two years later all private-sector employers with more than 10 employees will be required to register. All public-sector employers will be automatically registered. All registered employers must submit to the FEC annual monitoring returns giving the religious composition of their work-forces. This will include monitoring returns on job applications received for all public-sector employers and all private-sector employers with more than 250 employees. Smaller employers must maintain records of applications. Using the FEC's 'Code of Practice' employers must review their recruitment, training, and promotion practices at least once every three years to determine if there is 'fair participation' of Protestants and Catholics in their work-forces and to assess the need for affirmative action measures.

The FEC is given 'power to investigate the employment practice of any public or private sector employer in Northern Ireland at any time'. Usually such investigations will be linked to the annual monitoring exercise and to the reviews employers are required to conduct. The FEC can specify the setting of goals and timetables, and direct employers to take affirmative action measures:

in order to favour intentionally an underrepresented group the Commission can direct training at a particular place in Northern Ireland, or for a particular class of persons (provided that selection for training does not take place on the basis of religious belief or political opinion); and it can direct contacts by employers with specific schools in order to encourage a greater flow of applicants from an under-represented community. (DED, 1989: 9)

The Act outlaws direct discrimination and, for the first time, indirect discrimination. 'It makes it illegal to use job selection requirements or conditions which—though applied equally to both religious groups—have disparate or adverse impact on one group, are not justifiable irrespective of religious belief or political opinion, and are to the detriment of the individual concerned' (DED, 1989: 14). Failure to comply with the requirements of the new legislation and the directives of the FEC is backed up by a series of both criminal and economic sanctions. In particular, the FEC has the power to serve a 'notice of disqualification' on a defaulting employer. Such a notice will debar the employer from public-sector contracts and government grants.

A new Fair Employment Tribunal has been established. The FET will deal with individual cases of discrimination and has powers to award compensation up to the level of £30,000. In addition, the FET will hear appeals from employers against directions of the FEC. Again cash penalties up to £30,000 can be imposed on employers in breach of its 'orders of enforcement'. Regardless of the outcome of individual cases before the FET, the FEC can form an opinion as to whether or not the employer concerned has failed to afford equality of opportunity and has the power to pursue remedial action where required.

Finally, the Act will be reviewed in 1995 by the Central Community Relations Unit (CCRU). The CCRU is a high-level co-ordinating unit within the Northern Ireland Civil Service reporting directly to the Secretary of State for Northern Ireland. In addition, SACHR's duty to advise on discrimination law continues and SACHR will be independently involved in the evaluation exercise. We now turn to this question of evaluation.

## *Review and Evaluation*

The review and evaluation of the new legislation will be led, within government, by the CCRU as it is 'best placed to assess the extent to which the fair employment issue is informing and influencing all Government's social and economic policies in Northern Ireland' (DED, 1989: 16). Moreover, the government announced in March 1990 that 'equal opportunity proofing' of government policy-making and legislation had been introduced into the work of the Northern Ireland departments. Under this proofing system, all policy and legislative proposals 'will first be considered against a guidance checklist to ensure that they do not give rise to direct or indirect discrimination on the grounds of religion, gender or against the married [*sic*]'.

Announcing the policy, the minister responsible for fair employment, Richard Needham, indicated that civil servants will receive training on equal opportunity issues and that 'departments will adopt a positive and pro-active approach to ensure equality of opportunity in all policy making' (NIO, 1990).

More detail is needed on precisely how this 'proofing' system will work, but it has the potential to open up a major new dimension to the development of public policy in Northern Ireland. In the post-war period, and especially since the institution of Direct Rule in 1972, a major principle for the development of social policy has been that of 'parity' with policy, services, and standards with the rest of the UK. Significant areas of social policy are automatically applied to Northern Ireland as in the rest of the UK, e.g. social security policies and associated benefits. In other areas there is more debate and the possibility of discretion, although policy variations in Northern Ireland which involve additional expenditure incur close scrutiny and require agreement from the Treasury. Now, however, with the introduction of the equality of opportunity 'proofing' system a new guiding principle for policy is, it would seem, being introduced. For example, a clear departure from prevailing principles of economic policy in Britain since 1979— reduction of government intervention, labour market deregulation, and the enhancement of the market—to meet fair employment objectives is signalled in a recent Department of Economic Development statement on economic policy in Northern Ireland which declares that government

has a responsibility to target its industrial development, training and enterprise programmes to tackle the long term problems of the unemployed, to meet the needs of areas of high social and economic deprivation, and to ensure that everyone has a fair opportunity to compete for employment . . . In seeking to help areas of high unemployment and social deprivation, the Government recognises that, in addition to . . . [these] . . . integrated packages of assistance . . . special measures may be necessary to assist or rescue firms where the future development of these areas may be severely affected by closure. (DED, 1990: 30–2)

Therefore, instead of economic policy in Britain being applied in Northern Ireland, a policy almost at variance is to be pursued with the clear objective of enhancing fair employment. Ministers appointed to the Northern Ireland Office with ideological leanings favoured by the current Conservative government are likely to be rather surprised by circumstances and policies they may be asked to endorse and promote. It will be important to monitor how this 'proofing' of policy is applied.

The evaluation of the fair employment initiative is going to constitute a major area of public debate and government concern during the 1990s. What should the review encompass? The SACHR laid down one criterion for evaluation when it conducted its investigation of fair employment. It suggested that the target should be the reduction of the differential between Catholic and Protestant male unemployment from 2.5:1 to 1.5:1 in five years. This criterion was rejected by the government. The SACHR made this recommendation in the light of a labour turnover figure of 100,000 per year supplied by the Department of Economic Development. Such a bench-mark has some utility but it needs to be very carefully understood. A goal is being set for government in all its various job-related activities. Fair employment legislation can only ensure that existing and newly created jobs are fairly distributed, it cannot create jobs. Hence the legislation and the Fair Employment Commission cannot and should not be judged a success or failure on this bench-mark alone.

Secondly, the 100,000 figure itself requires close attention. This figure is likely to include male and female, part-time and full-time, employment, as well as movements in and out of training schemes. The proportion of the 100,000 movements which relate to full-time male jobs, which is where the main unemployment difference lies between Protestants and Catholics, is likely to be quite small. It is clear that recent structural shifts in employment have been towards part-time and female jobs and in the major expansion of government-sponsored training schemes. Finally, it is likely that semi-skilled and unskilled work will record higher turnover levels than better-paid and more secure employment. Much of the employment recorded in these job changes is not, therefore, primarily of the type required to make the Catholic employment profile more equitable compared to Protestants.

There is no doubt that for many, both in Northern Ireland and internationally (especially in the US), the unemployment differential will be the prime yardstick for the evaluation of the new policy and the government's resolve in tackling structural inequality. In one way, this is not unreasonable: the unemployment differential has been used as the key indicator of the difference in the employment circumstances of the two communities, and was used by the government as the justification for the strengthening of the legislation. However, it will be important that the evaluation takes into account the full context of the labour market. For example, as we noted earlier, a report from the Northern Ireland Economic Research Centre (Gudgin and Roper, 1990) suggested that reducing unemployment would require the generation of three new jobs to enable one person to leave the unemployment register. In addition, the long-term unemployed, particularly Catholics, often live in areas which pose severe problems for job-creation agencies in attracting investment. Flexibility and adaptability are of the essence in such circumstances (see Rolston and Tomlinson, 1988, for a critical review of government's policies in West Belfast).

It is likely that considerable further improvement will take place for those Catholics with good educational qualifications entering professional, technical, and administrative posts (see Chapters 2 and 4). Such a development could leave the substantially disadvantaged areas of West Belfast, Newry, Derry, etc. largely untouched. The unemployment differential, as a single key indicator, has the advantage of keeping the problems of these areas to the forefront; but it would be both unreasonable and an almost predetermined judgement of failure to use this measure as the *sole* indicator of the success or failure of the new fair employment initiative.

Alongside the general assessment of the full range of government's social and economic policies there will be a need to rigorously assess the specific components of the new legislation:

1. The FEC is well funded and staffed. It must be seen to be operating efficiently and effectively. Relevant indicators to measure the performance of the FEC, across the range of its activities, must be developed.

2. The establishment of the Fair Employment Tribunals and their role in responding to individual complaints of discrimination invites evaluation and comparison with the work of the FEA—which was previously responsible for the investigation of individual complaints.
3. Little is currently known about the Fair Employment Support Scheme. An evaluation of the scheme encompassing expenditure and quality of advice given by consultants should be developed.

Much could be gained in the evaluation by developing appropriate international comparisons. The Canadian employment equity policy is being reviewed in 1991 and the criteria developed there could be useful in the Northern Ireland context. In addition, US equal opportunity policy is under continuous scrutiny and judicial redefinition.

Setting out the parameters of the evaluation of the legislation is important. The timescale for the review is short: only five years from the passing of the legislation. In our view, only relatively small changes in the broad socio-economic profiles of the two communities can be expected. It is almost inevitable, therefore, that demands for the further strengthening of the legislation towards a 'quotas' policy will continue.

*Early Responses to the Legislation and Initial Problems for the FEC*

R. A. Butler described politics as 'the art of the possible'. From such a perspective the new Fair Employment Act is a remarkable achievement. Mrs Thatcher's Conservative governments were committed to *laissez-faire* rather than the *noblesse oblige* of traditional Tories. Her three administrations worked through significant deregulations of the labour market, business, commerce, and parts of the activities of the professions; they proved distinctly unsympathetic to government interventions in delivering equality of employment opportunity to minority groups. However, here in Northern Ireland we have what the government itself claims to be 'the most radical fair employment law ever passed by the United Kingdom Parliament'.

The new legislation decisively moves affirmative action from the previous essentially voluntary position to one where affirmative action is virtually mandatory on employers, where mechanisms are in place to police compliance, and where there are enforcement arrangements of some power. However, McCrudden, in Chapter 10, considers there are flaws which may prove substantial. In the event of a challenge to the FEC's directions, the interpretation of key terms and practices by the courts will be crucial. For example, the Act defines affirmative action as 'action designed to secure fair participation in employment by members of the Protestant, or members of the Roman Catholic, community in Northern Ireland'. The White Paper defined affirmative action as action to provide 'a more representative distribution in the workforce'. McCrudden argues that the courts may not take the view that 'fair participation' equals 'a more representative distribution'. The latter, he argues, was results-oriented while the former may be interpreted as merely seeking to guarantee fair procedures.

Whatever the flaws or inadequacies in the new legislation much will depend on how forcefully and effectively the new Fair Employment Commission, with twice the staff and funding of the old FEA, operates. It is perhaps significant that the review, to be conducted after five years, has been couched in very broad terms: 'to assess the extent to which the fair employment issue is informing and influencing all Government's social and economic policies in Northern Ireland'. This suggests that the government is not narrowing its review to merely the efficiency and effectiveness of the FEC but rather is casting the net much wider, notably in its statements on equal opportunity 'proofing', to include all aspects of government policies and practices which might have an effect on fair employment. In many ways this is at the heart of the new approach. Much of the new law is designed to ensure that employers are forced to make certain that their procedures and practices confirm to fair employment principles. With its affirmative action provisions, however, there is undoubtedly a significant new emphasis on measuring outcomes as a yardstick of success. It is the measurement of employment outcomes, including recruitment policies, that determines whether affirmative action should be instituted. For example, the Northern Ireland Civil Service may be said to have ensured that its procedures and practices conform to reasonable standards, but there remains a significant under-representation of Catholics at senior levels. Should the NICS institute affirmative action to try to resolve this problem? Should the NICS recruit personnel from outside the service, contrary to usual practice, in order to try to enhance the representation of Catholics in the upper-echelon positions (Osborne, 1990)?

In Canada, all Crown Corporations and federally regulated corporations with more than 100 employees must submit annual monitoring returns to Canada Employment and Immigration Commission (CEIC). The professionalism and sophistication of this operation is impressive. CEIC has a considerable advantage in that Statistics Canada conducts five-yearly censuses which are used to provide detailed labour availability estimates: estimates of the availability of designated groups in the workforce. The monitoring returns and labour availability estimates are then passed to the Canadian Human Rights Commission, which has had enforcement added to its traditional functions of dealing with individual cases of discrimination (Cormack and Osborne, 1990).

The Fair Employment legislation in Northern Ireland imposes a tremendous task on the FEC in dealing with the annual monitoring returns. The Canada Employment and Immigration Commission is dealing with around 450 employers on an annual basis. In Northern Ireland almost 2,000 firms were registered with the FEC in the first year of monitoring. These are employers with more than 25 employees. When monitoring is extended to firms with more than 10 employees in 1992 the estimate is that this will then involve more than 4,500 employers in the annual monitoring exercise. The FEC face this daunting task without the luxury of the quality of labour availability estimates on tap in Canada. Existing data sources in Northern Ireland, such as the Labour Force Survey or the Continuous Household Survey, do not permit adequate analysis in geographical or occupational terms for availability purposes. The population census scheduled for 1991 may well suffer from the same problems of enumeration and non-response to the religion question as in 1981 and, in any event, is unlikely to be available to the FEC until 1993. It should be noted that labour availability estimates based on the *existing* distribution of jobs already reflect a distribution heavily influenced by discrimination and disadvantage. Much more useful labour availability estimates than those based on existing 'stock' data (job profiles of the two communities as revealed in major surveys) will come from 'flow' data (e.g. data on the output of schools, colleges, universities, and training schemes). However, major sources of new entrants to the labour market, such as those from training schemes, further education, and higher education, are not currently monitored in terms of religion, although work is under way to obtain such 'flow' data by religion. As the new legislation comes into force the present inadequacies of availability data and the problem of how to interpret them may cause problems for the FEC regarding the advice it can give to employers and the judgements it can form on an individual employer's employment profiles. While the monitoring data undoubtedly represent a substantial addition to knowledge of the labour market, the precise form of information required from employers, under the regulations, actually masks some key aspects of employment patterns necessary for informed judgements to be made. For example, employers whose activities have multiple locations (e.g. banks, major high-street retailers, etc.) are only required to submit a single aggregated monitoring return, thereby precluding a ready assessment of individual plants/branches in terms of local labour markets.

## *Key Terms and Choices: The Debate over Policy*

The civil servants drafting the new legislation carefully tried to negotiate severe obstacles, some merely semantic, but some with major implications for the working of the legislation. Originally, as we have seen, they attempted to defuse opposition to their approach by semantically following the Canadian approach and labelling the area 'employment equity'. This proved unpopular and they were thrown back to negotiating their way through definitions of 'affirmative action', 'positive discrimination', 'goals and timetables', 'quotas' etc: all terms with high emotive content but often lacking clarity.

Affirmative action is the most basic term encompassing various aspects of state intervention in the delivery of equal employment opportunities. The term first appeared in the United States Civil Rights Act of 1964. Glazer, in reviewing the period since then, states:

Affirmative action in employment became controversial only when it went beyond the written language of the Civil Rights Act and the Executive Order, and began to require employers to hire or promote specific numbers of minority applicants or employees . . . 'Quotas' or 'goals and timetables' became the buzzwords of choice in disputes over the appropriate degree of 'affirmative action' . . . The expectation of colour blindness that was paramount in the mid-1960s has been replaced by policies mandating numerical requirements . . . whatever the term meant in the 1960s, since the 1970s affirmative action has come to mean quotas and goals and timetables. (1988: 102)

It was to avoid the association of affirmative action with quotas that Judge Abella coined the term 'employment equity', and for the same reason civil servants sought to use the term for legislation in Northern Ireland.

Despite perceptions to the contrary, quotas have rarely been applied in the United States. The Supreme Court has only endorsed them in cases where there was clear evidence of 'egregious discrimination'. Employment equity in Canada and fair employment in Northern Ireland have been defined as affirmative action incorporating goals and timetables but not quotas (Cormack and Osborne, 1990). The Fair Employment Commission is empowered to issue a notice about goals and timetables which 'shall specify the period or periods concerned and, in respect of any specified period, the progress that, in the opinion of the Commission, can reasonably be expected to be made towards fair participation by members of the community concerned' (HMSO, 1989b). The SACHR adopted much the same position in its recommendations on fair employment. It argued against 'reverse discrimination' where ' an employer gives preference to members of underprivileged or underrepresented groups in order to increase their numbers regardless of their relative qualifications or suitability' (SACHR, 1987: 70). However, the SACHR left this particular door ajar by suggesting that there 'is a case for reverse discrimination only where other affirmative action measures have been tried and found to have failed' (SACHR, 1987: 71). Edwards usefully distinguishes between 'positive action' and 'positive discrimination':

positive discrimination substitutes the group characteristics for morally relevant ones (such as need or merit) in distributive processes, positive action attempts to ensure the more effective application of morally relevant criteria in respect of groups with certain characteristics. What distinguishes positive discrimination therefore from other social policy practices such as selectivity . . . is that its beneficiaries are identified by criteria (ethnicity, religion, gender) that are morally irrelevant to the purposes for which the practice is applied: meeting need, allocating jobs on the principle of merit. (Edwards, 1989: 15)

The new legislation in Northern Ireland strongly endorses a series of 'positive action' requirements, but equally strongly rejects positive discrimination. However, it is debatable whether this position is based solely on principle or is significantly determined by the perceived political response from the Protestant community were positive discrimination to be embraced.

The opponents of affirmative action most usually base their case on two types of argument: (1) that the rights of individuals in non-target groups are infringed; and (2) that the operation of the market is ultimately sufficient to ensure fairness in employment. The Reagan-appointed Justice Scalia, in the last in a major series of US Supreme Court decisions on affirmative action, dissented in the case of a transportation agency's affirmative action plan to increase the numbers of women and minority employees. Scalia argued:

The only losers in the process are the Johnstons [the white male over whom a female was promoted] of the country for whom [civil rights law] has not been merely repealed but actually inverted. The irony is that these individuals—predominantly unknown, unaffluent, unorganized—suffer this injustice in the hands of a court fond of thinking itself the champion of the politically impotent. (*New York Times*, 26 Mar. 1987)

In a similar vein Edwards argues that 'the majority of the most
deprived people in Britain are white ethnics, and it is this single
fact more than any other that makes positive discrimination in
favour of ethnic minorities in pursuit of the needs principle
unjust' (Edwards, 1987: 201). Despite this, Edwards is, as we
noted, willing to entertain a series of 'positive action' measures
particularly if based on 'morally relevant' criteria such as need
and not simply on the basis of group membership, e.g.
ethnicity, religion, or gender.

However, discrimination, by its very nature, is based on the
application of a collective label, e.g. black, Catholic, female.
The individual is not primarily treated as an individual but as
an imputed member of a group (Lustgarten, 1987). This
suggests that the problem may well have to be treated using
collective rather than individual solutions. In this context, it is
worth noting that, in Northern Ireland, the differences between
the Protestant and Catholic communities are nothing like as
great as those between blacks and whites in the United States.
The recent report of the Committee on the Status of Black
Americans (Jaynes and Williams, 1989) makes depressing
reading. The progress made by blacks in earlier decades does
not appear to have been sustained through the 1970s and
1980s. The dissolution of the black American family, with the
majority of black children under 18 living in families that
include their mothers but not their fathers (compared to four in
every five white children living with both parents), and the
continuing substantial gap in educational attainments between
black and white children, are not problems faced in Northern
Ireland. As Farley (1988) has suggested, the single-parent
family may be something increasingly common in white communities, and that would apply in the UK as well, but there is
no evidence to suggest that single-parent families are more
common in Catholic as opposed to Protestant communities in
Northern Ireland. Neither is the difference in the levels of
educational attainment in the two Northern Irish communities
on the scale of the difference between blacks and whites in the
United States. It might be noted here that the Welfare State, for
all its inadequacies, has, by and large, prevented the levels of
degradation all too common in US cities. Hence, at present,

there is a less strong case to be made in Northern Ireland for positive discrimination than there is in the United States.

The second type of argument used against affirmative action policies is one based on a faith in the unfettered power of the market. Glazer (1988) argues that there was a rapid rise in the employment of blacks in firms subject to equal employment opportunities legislation after the adoption of the Civil Rights Act banning discrimination but before the much stronger versions of affirmative action were introduced. Having dealt with economic and educational discrimination, following on from the civil rights struggle, Glazer feels the market should have been left to operate. As with immigrant groups blacks would eventually take their place throughout American society. Somewhat more crudely in Canada, Block and Walker also advocated the power of market forces:

the institution of profit and loss can be a powerful barrier against the expression of discrimination in the labour market. Those who indulge such prejudices will have to pay higher salaries for a given quality of employees, or make do with workers of lesser skills at the same wage. In either case, losses in the competitive struggle will tend to ensue. (Block and Walker, 1985: 13)

Such naïve faith in market forces has been popular, on both sides of the Atlantic, in recent years. But sources from Adam Smith, whose name is frequently taken in vain (Winch, 1978), to modern political economists and other social scientists, state that non-economic forces at work in the market cannot be ignored (see e.g. Gouldner, 1970). For present purposes the most useful concept is Weber's 'social closure'. Parkin, following Weber, states:

By social closure Weber means the process by which social collectivities seek to maximize rewards by restricting access to resources and opportunities to a limited circle of eligibles. This entails the singling out of certain social or physical attributes as the justificatory basis for exclusion. Weber suggests that virtually any group attribute—race, language, social origin, religion—may be seized upon provided it can be used for 'the monopolization of specific, usually economic opportunities'. This monopolization is directed

against competitors who share some positive or negative characteristic; its purpose is always the closure of social and economic opportunities to outsiders. The nature of these exclusionary practices, and
the completeness of social closure, determine the general character of
the distributive system. (Parkin, 1979: 44)

Market forces are unlikely, of their own accord, to break
long-established patterns of social closure, especially of the
type experienced in Northern Ireland. To use a topical meta-
phor, discrimination in employment pollutes the social environment in Northern Ireland. Just as the government must
intervene to reduce and control the pollution of the physical
environment, so too must it intervene to eliminate one of the
major causes of the suspicions and antagonisms evident in the
relations between the two communities in Northern Ireland.

### The Context for Social Policy Development in Northern Ireland

Social policy development in Northern Ireland, especially since
Direct Rule, has tended to concentrate on achieving 'parity',
both of standards and policy, with the rest of the UK. This
resulted in a huge expansion of expenditure on housing in the
1970s, for example. Examples of where Northern Ireland has
failed to achieve parity, in the case of law enforcement and the
1971 Payments for Debt Act (now to be abolished), are regarded as unfortunate products of the continuing political and
security situation. For Northern Ireland to be innovatory in
'positive' policy developments is, therefore, rather unusual.
Although credit for these circumstances must lie largely with
the perceived political need to respond to the MacBride
campaign, there is also evidence that some NICS officials have
been advocates of extending policy and would appear to have
argued for a more comprehensive approach, based on the full
multi-dimensional model, than is currently being enacted.

There can be little doubt that the government's willingness
to initiate and implement a major strengthening of equal
opportunity policy in Northern Ireland owes much to the
international context of the Northern Ireland problem and
relatively little to its commitment to promoting equal opportunities. As the Northern Ireland proposals took shape the

phrase most often to be heard from civil servants was 'the problem of read-across', by which was meant the extreme sensitivity felt by the government to the argument that if such a strengthened policy could be promoted in Northern Ireland then why not in the rest of the UK? By confining the new policy to religion (and not gender and disability) it could be depicted as of relevance only to Northern Ireland—'a place apart'. However, for those advocacy groups in Britain who may look to the single market in the EC and the strengthening of the role of the European Commission to provide a stronger means than hitherto for the enhancement of equal opportunities policy, it may well be advisable to reject the 'place apart' thesis and to monitor closely the implementation of fair employment policy in Northern Ireland. In this the Commission for Racial Equality (CRE) has been quick to look to Northern Ireland and has called for the introduction of similar legislation in Britain (CRE, 1990).

## CONCLUSION

This chapter has ranged from broad comparative analyses to detailed, and often technical, examination of fair employment issues and policies. Ethnic divisions in modern societies often call for sophisticated and elaborate political structures in the attempt to hold such plural states together. The current problems in Canada following the 'Meech Lake' fiasco are but the most recent and striking example. Since the imposition of Direct Rule in Northern Ireland in 1972 various Secretaries of State have sought to find political structures which will be supported by both communities.

While the search for the elusive political solution must continue, some of the aspects of the divisions in Northern Ireland that create the need for a political solution in the first place can be tackled. Not least among these is fair employment. The British intelligentsia and media have notably failed to look beyond the political impasse and the latest atrocity. It has been too easy to describe Northern Ireland as a 'place apart' where ancient religious wars are re-enacted. Such a short-sighted view misses the contemporary relevance of understanding

Northern Ireland in the context of the increasingly visible and salient ethnic divisions in Europe and elsewhere.

The new fair employment legislation in Northern Ireland is well in advance of anything attempted currently in Europe. The goal of equal opportunities will now depend greatly on the rate of new job creation, the work of the Fair Employment Commission, the operation of the Fair Employment Tribunals, and, perhaps in particular, the degree to which the government pursues the fair employment 'proofing' of its policies. In the study of plural societies an understanding of the operation of the labour market and fair employment policies in Northern Ireland will undoubtedly repay those who search beyond the facile and inaccurate depiction of the Province as the location of a medieval holy war.

# 2

# Religion and the Labour Market: Patterns and Profiles

*Robert D. Osborne and Robert J. Cormack*

## INTRODUCTION

This chapter seeks to outline the labour market profiles of Protestants and Catholics in Northern Ireland. It examines the historical evidence before turning to more recent times and assessing economic activity, unemployment, occupations, and industrial distributions. The principal sources of data are the population censuses of 1971 and 1981 and the results now available from new continuous surveys, the Continuous Household Survey (CHS) and the Labour Force Survey (LFS). The chapter draws on previous analyses undertaken by the authors of these data sources (Osborne and Cormack, 1987; Osborne, 1987; Cormack, Osborne, and Curry, 1989).

The chapter also attempts to use these data sources, not only to outline contemporary patterns, but also to assess the extent of change in these profiles. The expectation of change derives principally from the existence, from 1976, of a fair employment policy which has sought to eliminate discrimination and promote equality of opportunity. However, other circumstances in Northern Ireland, alongside evidence from other countries, suggest that change is more likely to be incremental rather than substantial. The continuing civil disturbances have not only resulted in the physical destruction of jobs but have also acted as a deterrent to new investment and jobs (Simpson, 1983). Moreover, national economic policy in the 1980–1 period resulted in a major collapse in manufacturing employment (NIEC, 1981). Unemployment increased dramatically and has only begun to decline since the late 1980s. Employment in the public sector increased during the 1970s before stabilizing in the early 1980s. With the decline in manufacturing, employment in the service sector increased and the public-sector share of employment reached 45% by the end of the 1980s. Much of the new service-sector employment has been part-time and much of it taken up by women. If the particular economic and political circumstances have been unconducive to dramatic change in employment patterns, the experience of other countries with much stronger legislative and policy frameworks, most notably the US, further cautions against expectations of substantial change. Although some of the evidence is contradictory there is little to suggest that affirmative action policies in the US have substantially improved the socio-economic position of black citizens (Farley, 1988; Jaynes and Williams, 1989). For these reasons we could anticipate the scale of change to be incremental rather than fundamental. However, on a purely meritocratic basis, the evidence of the improving levels of educational attainment of Catholics (see Chapters 4 and 5) suggests we might expect some convergence in the profiles of newer entrants to the labour market.

## THE HISTORICAL EVIDENCE

The analysis of historical labour market differences between Protestants and Catholics has been sparked, to a significant extent, by contemporary academic and public policy concerns with the issue. Borrowing techniques from sociologists, historians have begun to analyse data from the early part of the twentieth century. Using the 1901 census, Hepburn and Collins (1981) provide a series of measures of religious segregation and industrial and occupational profiles of Protestants and Catholics for Belfast. Their analysis reveals a lower proportion of Catholics, based on heads of household, in manufacturing mainly because of low representation in engineering and shipbuilding. Higher figures are recorded in commercial occupations and shopkeeping, in general labouring, and marginally so in public services and the professions. Grouping occupations into social classes demonstrates, when compared with figures for Great Britain for 1911, that 'the Catholic population of Belfast did not occupy an inferior position in relation to Britain as a whole, but rather that the Protestant population had obtained for itself a relatively advantaged position' (Hepburn and Collins, 1981: 226).

Hepburn (1983), using a random sample of manuscripts from the 1901 census, and information from a sample of data from 1951 marriage register data, attempted to compare the socio-economic profiles of Protestants and Catholics and to assess rates of social mobility between the two dates. Despite some problems with the data, Hepburn's conclusions firmly point to a disadvantaged Catholic employment profile and lower rates of social mobility:

It is clear from our analysis that major differences exist in the employment profiles of Catholics and Protestants at the beginning of the century and that they did not, in general, narrow in the ensuing fifty years. The disadvantageous position of Catholics which had increasingly slipped during the second half of the nineteenth century was not alleviated during the first half of the twentieth . . . The individual analysis of intergenerational mobility by religion confirms this general picture. (Hepburn, 1983: 61–2)

This conclusion is supported by an analysis of social mobility drawing upon a major sample survey of adult males reported by Miller (1983). The sample, drawn during 1973–4, collected the employment histories of individuals together with details of intergenerational mobility and spanned a significant part of the twentieth century. Miller concludes:

In terms of the actual group experiences of mobility, the Catholic experience has been realised . . . in somewhat of a disadvantage in origin exacerbated by a widening and clarified disadvantage in the present . . . What the [above] results convincingly demonstrate is that it would be inaccurate and foolish to assume that present day inequities are only legacies from the past that will somehow gradually fade into oblivion of their own volition. (Miller, 1983: 76)

These studies reveal the extent to which there was a significant employment disadvantage experienced by Catholics at the turn of the twentieth century and show that there was little evidence of it declining over the next sixty years.

Prior to these studies some attempts had been made to assess the employment profiles of Protestants and Catholics. Barritt and Carter (1962) made some limited assessments of practices in the private sector. They suggested that four situations could be found: (i) firms which only employed individuals from one community; (ii) Protestant-owned firms which employed Catholics only, in lower-paid jobs, not in supervisory positions; (iii) firms employing both Protestants and Catholics but segregating them by departments; (iv) firms with a mix of members of the two communities within the same departments. They were unable to derive an estimate of the cumulative impact of these practices for labour market profiles. They did, however, note the considerable underrepresentation of Catholics in the Northern Ireland Civil Service and the widespread discrimination in local authority employment.

A book published soon after the outbreak of the 'troubles', based on survey data collected at the end of the 1960s, suggested that there was no discernable aggregate discrimination against Catholics in either public housing or employment (Rose, 1971). These conclusions conflict with the conclusions of most other observers of the period and may well stem from a sample response which was skewed towards middle-class respondents, especially in the case of Catholics. The official commission of inquiry, set up by the Unionist government to investigate the outbreak of civil disturbances at the end of the 1960s, the Cameron Commission, found that 'Social and economic grievances or abuses of power were in a very real sense an immediate and operative cause of the demonstrations and consequent disorders after 1968' (Cameron Commission, 1969).

The most thorough attempt to assess the general pattern of local and central government behaviour up to the end of the 1960s, in terms of employment and other areas, has been that of Whyte (1983). He concludes his wide-ranging assessment with a list, in order of demerit, where discrimination was most prevalent: electoral practices, public employment, policing, private employment, public housing, and regional policy. Whyte qualifies this list by suggesting that many of the sources of complaint stemmed from Unionist-controlled local authorities, especially in the west of the Province. Nevertheless, he concludes: 'The unionist government must bear its share of responsibility. It put through the original gerrymander which underpinned so many of the subsequent malpractices, and then, despite repeated protests did nothing to stop those malpractices continuing' (Whyte, 1983: 31).


## THE EVIDENCE FROM THE 1971 CENSUS

There is no doubt that much of the research analysing the historical characteristics of the labour market in Northern Ireland was sparked by the detailed analysis of the 1971 census data, the first cross-tabulation of census employment data by religion since 1911, reported by Aunger (1975). Aunger's analysis revealed, for the first time, the nature of the employment, occupational, and industrial differences between Protestants and Catholics. The analysis suggested four dimensions of inequality:

1. Skilled/ unskilled: on the basis of modal averages the 'typical' Protestant male was a skilled worker, and the 'typical' Catholic unskilled.
2. Masculine/feminine: occupations which could be identified as strongly Protestant tended to be male, while a significant number of those identifiable as disproportionately Catholic tended to be disproportionately female.
3. Superordination/subordination: many of the occupations involving levels of authority and influence tended to be dominated by Protestants, while many of the lower-status service occupations were disproportionately Catholic.
4. Employed/unemployed: although less than a third of the economically active population of Northern Ireland, Catholics constituted a majority of the unemployed.

Aunger's analysis further revealed the detail of the differences, as shown in Tables 2.1–2.3. Table 2.1 shows the occupational class profiles of economically active men and women. Protestants were disproportionately represented in non-manual and skilled manual occupations, while Catholics were disproportionately represented only in semi-skilled, unskilled, and unemployed classes. While occupational disadvantage was evident for Catholics, it was also clear that a Catholic

TABLE 2.1. *Religion and occupational class (males and females) 1971* (%)

| Occupational class | Catholic | Protestant |
|---|---|---|
| Professional, managerial | 12 | 15 |
| Lower grade, non-manual | 19 | 26 |
| Skilled manual | 17 | 19 |
| Semi-skilled manual | 27 | 25 |
| Unskilled, unemployed | 25 | 15 |
| TOTAL | 100 | 100 |

N = 564,682

*Notes*: Based on those stating a religious affiliation and who were economically active; occupation classified using the Hall–Jones system; farmers and armed services omitted.

*Source*: Aunger, 1975.

middle class existed, with 31% in non-manual occupations, compared to 41% of Protestants in these occupations. It was also apparent that the size of the middle class was substantially a product of meeting the needs of the Catholic community. For example, 34% of Catholics in the professional and managerial categories were teachers (in separate Catholic schools) and clergymen compared with 19% of Protestants. On the other hand, the Protestant profile showed a much wider distribution

TABLE 2.2. *Religion and occupational class (males), 1971* (%)

| Occupational class | Catholic | Protestant |
|---|---|---|
| Professional, managerial | 9 | 16 |
| Lower grade, non-manual | 12 | 17 |
| Skilled manual | 23 | 27 |
| Semi-skilled manual | 25 | 24 |
| Unskilled, unemployed | 31 | 16 |
| TOTAL | 100 | 100 |

N = 365,948

*Notes*: Farmers, members of the armed services, and the self-employed omitted. Occupation classified using the Hall–Jones system.

*Source*: Aunger, 1975.

in administration, finance, and business services. These differences were even more marked for males, emphasising the disadvantaged position of Catholic males (Table 2.2). Aunger also observed that when manual occupations were considered there was a marked tendency for Catholics to predominate in lower-status occupations in industries both where they constituted a minority and where they constituted a majority. This can be seen in Table 2.3 where Catholics recorded higher proportions in low-status occupations in engineering (which had a low representation of Catholics) and in construction (which accounted for 40% of Catholic male employment in 1971).

Aunger's analysis did much to stimulate further work on aspects of the labour market profiles of the two communities. This was further added to by the creation of the Fair Employment Agency in 1976. The FEA began to conduct and sponsor its own research and with the availability of new data sources it is now possible to assess the extent to which employment patterns have changed since the bench-mark established by Aunger.

TABLE 2.3. *Religion and selected occupations in construction and engineering,*
*1971 (%)*

| Occupation | Occupational class | Catholic |
|---|---|---|
| Construction | | |
| Managers, building and contracting | I | 18 |
| Carpenters and joiners | III | 35 |
| Bricklayers, tile setters | III | 51 |
| Plasterers, cement finishers | III | 51 |
| Labourers, building and construction | V | 55 |
| Engineering | | |
| Managers, engineering | I | 8 |
| Fitters | III | 15 |
| Electricians | III | 20 |
| Motor mechanics | III | 27 |
| Labourers, engineering | V | 16 |

*Source*: Aunger, 1975.

## ECONOMIC ACTIVITY

The economic activity rate measures the proportion of a given population who are active in the labour market, being either in employment (part-time or full-time) or unemployed. Economic activity rates for the mid-1980s are shown in Table 2.4. The data are drawn from the 1985 Labour Force Survey and show a lower economic activity rate for Catholics at 55.2% than for Protestants at 59.9%. Most of this difference is attributable to the different activity rates for men: the Protestant male economic activity rate is 75.7% and the Catholic male economic activity rate is 70.8%. These activity-rate differences are evident across all age categories and even extend to those beyond the normal retirement ages. In relation to marital status there are quite striking activity-rate differences between married Protestant and Catholic females. Protestant married females are more likely to be economically active and this seems to be related to a significantly higher proportion of married Catholic females having young children into their early forties. Economic activity rates for married Catholic women only begin to match those of Protestant women for the 45+ age groups. These different activity rates have not been subjected to detailed assessment, but lower Catholic male rates may well be a product of the experience of high unemployment over many years. An assessment for 16–24-year-olds based on LFS data has suggested this reason (Osborne and Cormack, 1989*b*).

TABLE 2.4. *Economic activity rates by religion and gender, 1985* (%)

|  | Catholics | | | Protestants | | |
| --- | --- | --- | --- | --- | --- | --- |
|  | Male | Female | All | Male | Female | All |
| Economically active | 70.8 | 40.9 | 55.2 | 75.7 | 45.6 | 59.9 |
| Economically inactive | 29.2 | 59.1 | 44.8 | 24.3 | 54.4 | 40.1 |
| TOTAL | 100 | 100 | 100 | 100 | 100 | 100 |

*Note*: Based on those aged 16+.

*Source*: Cormack, Osborne, and Curry, 1989.

UNEMPLOYMENT

The 1971 census revealed that unemployment was substantially higher for Catholics than Protestants. This difference has largely remained the same from that date until the end of the 1980s, as Table 2.5 reveals. The data also reveal that the scale of the difference is much greater for males than females. As unemployment began to increase markedly from the end of the 1970s there was a widespread feeling that the differential experience of unemployment was being eroded: jobs were being lost in manufacturing and in areas of Northern Ireland which were predominantly Protestant. What was increasingly happening was a new common experience of high unemployment—an 'equality of misery' (Cormack and Osborne, 1985). One study, however, involving the secondary analysis of a government survey, conducted a detailed assessment of the experiences of 3,500 males monitored for a year after becoming unemployed. The study found major differences in the experiences of Protestants and Catholics:

Catholics were over-represented in the survey (compared with their representation in the economically active population), were more likely to have been unemployed in the previous three years, for that unemployment to have lasted longer, to experience a longer period before securing a job (for those obtaining a job), to receive from employment offices fewer job submissions, and to be disproportionately represented amongst those unemployed throughout the

TABLE 2.5. *Unemployment and religion* (%)

|  | Protestant | | | Catholic | | |
|---|---|---|---|---|---|---|
|  | Male | Female | All | Male | Female | All |
| 1971 census | 6.6 | 3.6 | 5.6 | 17.3 | 7.0 | 13.9 |
| 1981 census | 12.4 | 9.6 | 11.4 | 30.2 | 17.1 | 25.5 |
| 1983–4 CHS | 15.0 | 11.0 | 13.0 | 35.0 | 17.0 | 28.0 |
| 1985 LFS | 11.9 | 10.3 | 11.3 | 30.8 | 16.2 | 25.5 |
| 1985–6 CHS | 14.0 | 9.0 | 12.0 | 36.0 | 15.0 | 27.0 |

*Note*: CHS = Continuous Household Survey; LFS = Labour Force Survey.

*Source*: Constructed from census and survey reports.

year. Moreover, the analysis demonstrated that these differences could not be accounted for by variations in education, skill level (as measured by social class), geographical mobility or generational motivation. (Miller and Osborne, 1983: 98).

This study was also able to assess the role of Employment Officers in submitting individuals to job openings. What emerged was that, when other variables were controlled for, the religion of the individual was associated with the number of job submissions. This implied that the religion of individuals was being taken account of in the job submission process. This could have been in the form of submitting 'horses for courses', or it could have been that there was direct discrimination, or that both practices were taking place. By the time the data were analysed Job Centres were replacing Employment Offices and self nomination was replacing submission by Employment Officers. However, this study may have found evidence, often alleged anecdotally, that religion was a factor taken into account in official interventions in the labour market.

In the 1980s new data became available which cross-tabulated religion with unemployment: in particular, the 1981 census, the new Continuous Household survey, and the Labour Force Survey. Unemployment rates from these sources are shown in Table 2.5. It can be clearly seen that the differential in rates between Protestants and Catholics has remained fairly static during the 1980s and, indeed, shows little change from the early 1970s. Further analysis reveals that Catholic unemployment rates were higher throughout Northern Ireland, a higher proportion of Catholics were unemployed for over a year, and, Catholic unemployment rates were higher for all age categories (Osborne and Cormack, 1986).

There can be no doubt that there is little evidence of a convergence in the experience of unemployment between the two communities: there is no 'equality of misery'. Protestant unemployment rates increased markedly with the recession in manufacturing in the early 1980s but were partially offset by the expansion of security-related employment (see below). For Catholics, starting from both a much higher level of unemployment and a much longer experience of high unemployment, the

TABLE 2.6. *Religion and social class (unemployed included), 1981 (%)*

| Social class | Protestant | | | Catholic | | |
|---|---|---|---|---|---|---|
| | Male | Female | Total | Male | Female | Total |
| I | 4.5 | 1.0 | 3.2 | 2.8 | 0.8 | 2.1 |
| II | 22.9 | 20.1 | 21.8 | 17.4 | 23.7 | 19.7 |
| III Non-manual | 11.9 | 34.8 | 20.6 | 6.8 | 24.3 | 13.2 |
| III Manual | 29.5 | 6.6 | 20.8 | 25.2 | 6.8 | 18.5 |
| IV | 13.1 | 21.9 | 16.5 | 10.9 | 22.1 | 15.0 |
| V | 4.9 | 5.7 | 5.2 | 5.6 | 4.6 | 5.2 |
| Unemployed | 13.1 | 9.9 | 11.9 | 31.4 | 17.7 | 26.4 |
| N | 210,615 | 129,368 | 339,983 | 99,143 | 56,982 | 156,125 |

*Source*: Osborne and Cormack, 1987.

reductions in public expenditure during the 1980s severely hit
capital projects and thereby caused major problems in the
construction industry, a substantial area of employment for
Catholic males (see below).


## SOCIAL CLASS

In Table 2.6 we have constructed social class profiles from the
1981 census. From this table it is apparent that a higher
proportion of Catholics are in manual and unemployed categories (65.1%) compared to Protestants (54.4%), and that
Catholics show a lower representation amongst non-manual
groups especially in social class III non-manual. However, to a
significant extent, the Catholic profile is overshadowed by
unemployment: once only those in employment are compared
the profiles move much closer together (Table 2.7). In this
situation the modal category for both Protestant and Catholic
males in social class III manual with the second most significant category being social class II for both groups. The most
notable difference lies in the proportions of Protestants in social
class II, which is significantly higher than the proportion of
Catholics. Comparing the female profiles reveals comparatively small differences between Protestants and Catholics
albeit in the general context of the disadvantaged position of
females in the labour market.


## OCCUPATIONS

We can examine in more detail the extent to which change is
taking place by examining individual occupational categories.
There is some evidence of change in the managerial and
supervisory categories. Thus, while the total numbers recorded
as 'managers in large establishments' increased from 1971 to
1981 by 145%, Catholic numbers increased by 244% and
Protestant numbers by 55%. For all managers, Catholics
represented 16% of the total in 1971 and 18.5% in 1981.
Similarly, when supervisory occupations are considered, some
Catholic advance is recorded, especially for non-manual supervisory jobs, although most of this latter change is confined to

TABLE 2.7. *Religion and social class (unemployed excluded), 1981 (%)*

| Social class | Protestant | | | Catholic | | |
|---|---|---|---|---|---|---|
| | Male | Female | Total | Male | Female | Total |
| I | 5.2 | 1.1 | 3.6 | 4.1 | 0.9 | 2.8 |
| II | 26.3 | 22.3 | 24.8 | 25.3 | 28.8 | 26.7 |
| III Non-manual | 13.8 | 38.6 | 23.4 | 9.9 | 29.5 | 17.9 |
| III Manual | 34.0 | 7.3 | 23.6 | 36.7 | 8.3 | 25.1 |
| IV | 15.1 | 24.3 | 18.7 | 15.8 | 26.9 | 20.4 |
| V | 5.6 | 6.4 | 5.9 | 8.2 | 5.6 | 7.1 |
| N | 183,039 | 116,593 | 299,632 | 68,049 | 46,877 | 114,926 |

*Source*: Osborne and Cormack, 1987.

females. Further assessment of these jobs reveals that Catholics tend to record a higher representation in occupations of lower economic or strategic importance. For example, Catholics represent 40% of those managers in hotels, clubs, entertainment, and sport, 25.3% in retailing and wholesaling, but only 14.5% of office managers and 15.9% of production, works, maintenance, etc. managers. Moreover, when younger age groups are examined there is no discernible change in this pattern.

Within other occupational groups the following characteristics can be observed.

1. Catholics show a high representation in professionals in education, health, and welfare and in particular as teachers and nurses (especially females). Within this occupational group Catholics are not as well represented amongst doctors and dentists, pharmacists, radiographers, and other paramedicals and higher education teachers. Although Catholics tend to record lower proportions in higher-status jobs in this category there is clear evidence that Catholics, both male and female, show rising proportions in this group of occupations.

2. Construction occupations record a very high proportion of Catholics (48%) and there is no evidence that this characteristic is declining within younger age groups.

3. Probably the most important strategic occupational group is professional and related support management (including senior national and local government managers). Within this group there are substantial variations in individual occupations. Catholics account for 37.8% of judges, barristers, advocates, and solicitors but only 20.5% of economists, statisticians, and computer programmers, 14% of those in marketing, sales, advertising, and public relations, 15.9% of those recorded as general administrators—national government, 17.8% of senior local government officers, and 15.6% of those recorded as personnel and industrial relations managers. An increase is apparent, however, in these occupations amongst younger age groups.

4. Finally, in Aunger's detailed examination of the 1971 census, he noted that the occupations which recorded the highest proportions of Catholics were also predominantly

TABLE 2.8. *'Catholic' and 'Protestant' occupations, 1971 and 1981* (%)

| | Total employed (N) | Women | Catholic |
|---|---|---|---|
| **1971** | | | |
| **'Catholic' occupations** | | | |
| Publicans, innkeepers | 2,026 | 21 | 73 |
| Waiters, waitresses | 2,145 | 84 | 50 |
| Hairdressers, manicurists | 2,828 | 76 | 49 |
| Domestic housekeepers | 1,582 | 100 | 48 |
| Nurses | 12,249 | 90 | 43 |
| Primary, secondary teachers | 15,726 | 63 | 39 |
| **'Protestant' occupations** | | | |
| Company secretaries | 347 | 15 | 7 |
| Police officers and men | 4,046 | 3 | 10 |
| Chemists, biologists | 711 | 11 | 11 |
| Engineers | 3,282 | 0 | 11 |
| Managers | 10,312 | 6 | 12 |
| Senior government officials | 1,383 | 10 | 13 |
| **1981** | | | |
| **'Catholic' occupations** | | | |
| Hairdressers, barbers | 1,432 | 91 | 44 |
| Waiters, bar staff | 2,589 | 59 | 44 |
| Nurse administrators, nurses | 15,617 | 92 | 42 |
| Teachers | 18,226 | 63 | 41 |
| Managers in hotels, clubs, etc. | 2,589 | 34 | 40 |
| Judges, barristers, advocates | 940 | 11 | 38 |
| Catering supervisors | 1,483 | 66 | 38 |
| **'Protestant' occupations** | | | |
| Policemen, firemen, prison officers | 4,731 | 8 | 7 |
| Senior police, prison, and fire service officers | 439 | 4 | 10 |
| Engineers | 3,206 | 2 | 16 |
| Senior government officials | 2,179 | 16 | 16 |
| Managers (excl. retail, wholesale, and distribution) | 13,693 | 12 | 17 |
| Scientists, physicists, and mathematicians | 1,263 | 19 | 18 |
| Draughtsmen | 957 | 3 | 11 |

*Sources*: 1971 data adapted from Aunger, 1975; 1981 data from Osborne and Cormack, 1987.

female occupations and of lower status whereas those occupations which recorded the highest representation of Protestants were, in contrast, predominantly male and of higher strategic or economic status. The data are shown in Table 2.8. As can be seen, while the Protestant occupations have recorded little change, three of the Catholic occupations are of higher status—legal occupations, catering supervisors, and managers in hotels and clubs.

## INDUSTRIES

The census also allows an assessment of profiles in industries. The Catholic profile shows a high dependence on construction, an industry which consistently recorded higher numbers unemployed than employed during the 1990s. Both communities benefited from the expansion of public-sector jobs during the 1970s, but not in the same ways. For example, Protestants record a high representation in employment in security, which expanded considerably during the 1970s. In this way Protestants were able to offset the decline in manufacturing. (Most of this expansion was in the UDR, RUC, and prison service, which are areas Catholics have shown a traditional reluctance to enter, and those Catholics who do are often subject to intimidation by republican paramilitary groups.) On the other hand, Catholics tended to do better in the expansion of service-sector jobs. However, many of these were poorly paid and part-time and often taken by women. Protestants are generally strongly represented in manufacturing industries, especially aircraft and shipbuilding, whereas Catholics have a strong representation in education and professional services. The representation in agriculture is similar for both groups.

## THE EVIDENCE FROM FEA INVESTIGATIONS

The investigatory work of the FEA under Section 12 of the Fair Employment (Northern Ireland) Act 1976 (see Chapters 8 and 9) provides another source of information on employment profiles at the level of individual organizations. The position of the NICS is examined in Chapter 7 and the present authors reviewed the evidence from the most significant investigations up to 1986 (Osborne and Cormack, 1987). In general, for public-sector bodies there was evidence of increasing representation of Catholics but typically this was evident for recent years and Catholics were clustered at the lower levels of employment, e.g. the Southern Health and Social Service Board, with under-representation at more senior levels. Other organizations tended to show a much better pattern throughout the occupational profile, e.g. the Northern Ireland Housing Executive. On the other hand, the Electricity Service showed a generally poor representation of Catholics and little evidence of change and the Northern Ireland Fire Service showed a very poor representation of Catholics. In the public sector, an indication of the substantial change that has taken place since the original complaints of civil rights groups was that at the end of the 1980s six of the ten chief officers of the bodies responsible for the administration of health, welfare, education, and housing were Catholic. In the private sector there was some evidence amongst the banks of there being a policy of 'horses for courses', that is of sending Catholic staff to Catholic areas etc. There was a general under-representation of Catholics in insurance and building societies, although some change was apparent.

Two of the most significant of the recent investigations reported by the FEA were those completed into Northern Ireland's two universities: Queen's University, Belfast, and the University of Ulster. The evidence revealed by these investigations, and some of the responses to them, reveal why the new fair employment legislation has been required.

## QUEEN'S UNIVERSITY, BELFAST, AND THE UNIVERSITY OF ULSTER

The Queen's University of Belfast was constituted as an independent university in 1908 and, until the late 1960s, was the only university in Northern Ireland. A second university, the New University of Ulster (NUU), was opened in Coleraine in 1968 and a polytechnic, subsequently called the Ulster Polytechnic (UP), was opened on the northern outskirts of Belfast

TABLE 2.9. *Religious representation of employees at Queen's University, Belfast*

| Staff category | Protestant (%) | Catholic (%) | N | Non-Northern Ireland and unassigned (N) | Total |
|---|---|---|---|---|---|
| Academic | 82.0 | 18.0 | 411 | 382 | 793 |
| Administration | 89.2 | 10.8 | 111 | 27 | 138 |
| Technical | 78.5 | 21.5 | 381 | 45 | 426 |
| Clerical | 80.6 | 19.4 | 355 | 37 | 392 |
| Computer | 64.5 | 35.5 | 121 | 14 | 135 |
| Miscellaneous | 80.2 | 19.8 | 587 | 44 | 631 |
| Research | 74.0 | 26.0 | 215 | 83 | 298 |
| Library | 73.0 | 27.0 | 89 | 27 | 116 |
| TOTAL | 79.0 | 21.0 | 2,270 | 659 | 2,929 |

*Source:* Osborne and Cormack, 1990.

shortly afterwards. The NUU and UP were merged to form the multi-campus University of Ulster (UU) in 1984.

The FEA's assessment of Queen's commenced on the basis of employment data as of January 1987. The basic evidence is shown in Table 2.9. In contrast to many other employers in Northern Ireland, Queen's draws many employees from outside Northern Ireland. This, as can be seen from the table, is especially so for academics and researchers but much less so for other employment categories. In the presentation of data the FEA was careful to calculate proportions in terms both of locally recruited personnel and of all employees. The FEA was also careful to point out that even for locally recruited personnel some jobs at Queen's draw upon a catchment which extends throughout Northern Ireland while others have a more restricted geographical basis. Even taking all these important qualifications into account the picture that emerges is of a low representation of Catholics amongst those recruited from within Northern Ireland for academic and non-academic jobs alike. For example, only 10.8% of locally recruited administrative staff are Catholic, 19.4% of clerical staff, 19.8% of academics and researchers, and 19.8% of technical employees. On the other hand, 27% of library staff and 35.5% of those in computing are Catholic. Significantly, amongst these latter two groups, Catholics had been recruited in more recent times.

Quite marked differences were evident amongst academic staff by faculty. Even faculties which had a high proportion of staff from Northern Ireland such as Medicine and Engineering recorded proportions of Catholics of around 20%. Amongst the Miscellaneous category a very low figure was recorded for boilermen (7.3%), and even cleaning staff only recorded a figure of 15.9%.

In conclusion the FEA stated:

Overall the Agency found that there were major areas of under-representation of Roman Catholics of Northern Ireland origin relative to Protestants of Northern Ireland origin. The agency would expect the primary research and educational institution [*sic*] of the Province to take a leading role in matters of such importance as the promotion of equality of opportunity. However, until the time of this investigation, the University had not monitored its own employment pattern, it had not reviewed the effectiveness of its recruitment procedures nor had it assessed the impact of its actions on the relative opportunities for employment offered to Protestants and Roman Catholics. (FEA, 1989*a*: 43–5).

The report on the University of Ulster was released after the Queen's report, by which time the FEA had been redesignated the Fair Employment Commission (FEC) following the passing of the 1989 legislation (see Chapter 10). The FEC looked somewhat more favourably on UU than the FEA had on Queen's. As with Queen's, academic and research staff are only a minority of employees. The proportion of all staff locally recruited is higher in UU than in Queen's: 80.3% to 77.5%; while for academics the figures are 51.8% at Queen's and 58.4% at UU. Taking only these locally recruited staff, Catholics at UU represent 25.6% of academic staff, 43.1% of research staff, 25.7% of administrative and related staff, 20.5% of technicians, but only 12.0% of senior library staff, 14.8% of manual workers, and 11.1% of maintenance staff. Somewhat surprisingly, Catholics composed only 5% of computer staff, a category in which they are well represented in Queen's.

The FEC concluded in its report on UU: 'it is clear that significant change has taken place. The proportion of Roman Catholic applicants and appointees to posts as a whole in the University has exceeded 30%. The Commission welcomes the commitment and cooperation shown by the University' (FEC, 1990: 45). The University began monitoring recruitment during 1986 and this, together with an improving profile at the UU, seems to have led the FEC to be more favourably disposed to the UU than to Queen's.

However, it should be noted that, following the merger of NUU and UP in 1984, the number of academics at the UU increased by 15% from 757 in 1984 to 873 in 1988, while during the same period there was a slight decline in numbers at Queen's; a factor not mentioned in the FEC report. At the same time, the development of the Magee campus of the University in Derry, given the much higher representation of Catholics in its hinterland compared to either Coleraine or Belfast, helped enhance the proportion of Catholics in the University as a whole.

Nevertheless, the most remarkable feature of the position of both universities in these investigations lies in the absence of any action by either of them to implement the 1978 *Guide to Manpower Policy and Practice* (DMS, 1978) or the revised version of 1987 until faced with a formal investigation by the FEA. Both institutions advanced the line that, since their Charters outlawed religious discrimination and they had signed the FEA's Declaration, they had done all that was required of them. In this they adopted a position similar to that of many employers in the first years after the creation of the FEA in 1976 (see Chapter 6). However, it required a particularly well-developed myopia to ignore the increasing adoption of monitoring procedures by many employers following FEA investigations, including the NICS, and the clear evidence from 1986 that government was in the process of substantially reforming the fair employment legislation. A number of factors probably underpin the failure to respond to these developments. First, there has been a desire to keep the universities 'above' the conflict in Northern Ireland, as places of security and sanity untainted by the 'sordid' characteristics of the local conflict (Taylor, 1988). Secondly, there has been a preoccupation with the idea that universities are staffed by academics, despite the fact that academics represent only around a third of employees. Thirdly, there is a prevalent notion that academics are intrinsically incapable of malpractice. Finally, both universities, as the investigations revealed, have very few individuals drawn from the Catholic community, who might be most sensitive and receptive to these issues, at the more senior levels (Osborne and Cormack, 1990).

## CONCLUSIONS

The labour market patterns we have discussed clearly have their origins in the past going back to the early industrialization of north-east Ireland. The evidence suggests that these patterns were well entrenched by the early part of the twentieth century and that they have been replicated and reproduced over many years. They have persisted through major changes in the structure of the Northern Ireland economy. For example, they were maintained through the period of substantial investment in Northern Ireland, by external companies, during the 1960s and early 1970s.

However, we have also revealed that some changes are under way. There is some clear evidence that Catholics are beginning to move into some areas of management and supervisory occupations and that a range of professional occupations show increasing Catholic representation, and this is especially marked for those in the younger age groups. To a significant extent, the Catholic middle class is changing in composition and probably increasing in size. The absence of evidence of discrimination in the early careers of Catholic graduates (see Chapter 4) reinforces this evidence. Expansion into public-sector professional posts and other similar occupations means that the dominance of occupations related to servicing the Catholic community is being reduced. Although we have no direct data on this, it is quite likely that there may also be a geographical dispersal of the Catholic community, especially from West Belfast to suburban areas of Belfast and the commuter/market towns up to fifteen miles from Belfast. The geographical spread of these younger professionals will be governed to some extent by various issues including the accessibility of Catholic schools (see Chapter 4). What the social and political consequences of these developments will be remains to be examined.

The position in relation to manual occupations and the unemployed shows little evidence of change. The much larger experience of unemployment in the Catholic community, especially for males, has remained at a level at least twice that for Protestant males. Not a great deal is known about the contemporary stock and flow of individuals who are unemployed. However, it is clear that not only is the overall rate of unemployment higher for Catholics but this experience extends to both younger and older groups, the duration of unemployment is significantly longer for Catholics, and some Catholic areas have recorded very high levels of unemployment over decades. As the evidence suggests, even those young people from such areas who achieve modest success in terms of educational qualifications or who participate on the government training schemes, do less well in the labour market than their Protestant peers. This suggests the continuing reproduction of these circumstances despite government commitment to the provision of fair employment.

The changes that may be developing could be leading to a greater social differentiation between, on the one hand, those better-educated Catholics who are increasingly entering professional, technical, administrative, managerial, and supervisory jobs in both the private and public sectors, and the unqualified and unemployed on the other. It is possible that the 1990s will see a continuing expansion and occupational diversification of the Catholic middle class and a continuing marginalization of the socially, educationally, and economically disadvantaged, geographically concentrated in areas of West Belfast, Derry, Newry, etc. These differences seem likely to continue to be more marked for males than for females.

# 3

# Demography and Unemployment in Northern Ireland

*David Eversley*

## INTRODUCTION

In the debates which have raged around the 'problem of Northern Ireland' for generations, but particularly since the conflict between the communities escalated at the beginning of the 1970s, few statistical assertions have remained un-challenged. One that is not seriously disputed, however, is this: unemployment amongst Catholics is twice as high as amongst 'others', i.e. mostly Protestants (Cormack and Osborne, 1983; Osborne and Cormack, 1986). This has been so regardless of the general level of unemployment, which rose from a low of 6% to over 20% in the 1980s.

The main political debate has been about the causes of this differential. In its crudest form, the argument of one side has been that it is due to discrimination practised by Protestant employers against Catholic workers; on the other side we have the assertion that failure to obtain employment can be blamed largely on the Catholics themselves—mainly because their fertility is too high and they thus contribute too many entrants to the labour market (Compton, 1981).

This latter assertion is then usually qualified: not only are there too many Catholic children, but they also do not possess the necessary qualifications to take up the jobs which are offered, they live in areas where there are no opportunities, and they are unwilling to move to places where there is work.

The pattern of explanations is familiar to students of the economic problems of minorities elsewhere in the world. In the North American labour market in particular, excess black unemployment in the United States has frequently been 'ex-plained' in almost precisely the same terms as those used in

Northern Ireland. It is important to note that although the actual situations are quite different, the *explanations* tend to be similar; thus suggesting that the key to the problem lies in the nature of the political conflict rather than in any objective analysis of the underlying facts.

There is an important difference, however. Whereas explanations blaming the minorities themselves were common until the Kennedy/Johnson era, they gradually began to become unacceptable. References to US black fertility, or lack of education, or concentration of black families in areas of low economic growth, are now heard only from local politicians anxious to avoid paying for programmes which would change the situation, and to keep the black population in a subordinate condition, even when civil rights have been granted. At the national level, 'blaming the victim' went out of fashion.

Instead, programmes were initiated to remedy the skill deficiencies of some of the unemployed workers: overt acknowledgement that what had been mainly at fault was a political system which denied opportunities to blacks to become integrated into the labour force. Investigators began to concentrate on the root causes of black poverty: why was it that blacks received less education than whites? Why did they stay in the slums of Alabama and Mississippi and Harlem? And why did employers fail to offer them employment even when labour was in short supply?

Although the 'American Dilemma' (Myrdal, 1944) has not by any means been resolved, it is striking to compare the current approaches in the two countries: in Northern Ireland the recognition that the remedies for the problem are largely a matter for political action is of much more recent date than in the USA. In fact, in the United States, the debate about the causes of different sorts of differentials between the majority and the minority populations has become much less extreme. There are far more in-depth statistical analyses (Johnson, 1985–6). There are also now clear voices from the side of the minorities calling for self-help and self-improvement efforts to complement the relatively large expenditure on relevant programmes and the public attempts to ban discrimination (Lourie, 1984; Moynihan, 1986; Novak, 1987). In other words, the debate has become less polarized, more relevant to the perceived facts of the case, and therefore more constructive.

This juxtaposition is not an attempt to equate the position of Catholics in Northern Ireland with that of blacks in the USA— or Algerians in France, or Turks in other European countries. Such attempts have been made but they are often politically naïve and throw little light on the reality of prevailing conditions (Castles and Kosack, 1973). Rather, one points to the dissimilarities in the way the problem is frequently presented in order to draw attention to the fundamental differences. In the US, explanations of the 'racialist' type cited have been ruled out of court along with other forms of prejudice, whether relating to ethnicity, gender, or other personal characteristics. In Northern Ireland, however, it is still apparently considered respectable, even for members of the academic community, to absolve government and the majority community from blame for the condition of the minority population; and to infer that their troubles are of their own making. (Other specific forms of prejudice can also be shown to survive intact in Northern Ireland longer than in most other industrialized countries.)

It is not the purpose of this chapter to analyse the causes of the persistence of an attitude which is now no longer permissable elsewhere. Instead, we need to re-examine the facts to see why the idea that the Catholics are the authors of their own misfortunes is so persistent.

## DEMOGRAPHIC TIME-SCALES

Before we begin to analyse the evolution of the present demographic structure of Northern Ireland in relation to the labour market, we first have to be clear about some characteristics of the time-scales of the demographic and socio-economic interaction process.

First, we need to look at the framework in which the labour market operates. The demand side reflects economic conjunctures, mostly the present volume of production, but partly also expectations about the future, in Northern Ireland and elsewhere. Long-term changes in the composition of the national product, and shorter-term cyclical fluctuations, play a role. The supply side is a reflection of decisions to marry, or to reproduce, taken (however unconsciously) twenty to forty years earlier. It is also a reflection of past migration patterns and, to a much lesser extent, of mortality experience. It is a rather facile generalization in economic writings to assert that supply and demand are interdependent. In demographic analysis, this cannot be the case.

In 1986, the largest quinquennial age group was that in the range of 15–19 years, i.e. those born between 1966 and 1971. The median age group of males in the labour market in 1986 (this supply total is here assumed to consist of all males aged 20–59) was in the 35–39 year range. (Although there is also a problem of women's unemployment, for the sake of simplicity we will here address the male problem only—other factors, such as the propensity of married women to become or remain economically active, require a different sort of analysis.)

### THE DECISION TO PRODUCE THE SUPPLY OF LABOUR

The 'decision' to produce this supply of human capital was taken in the 1950s and 1960s. We could suppose for the purpose of this analysis that reproduction was a conscious act of will when these boys were born; it would be technically better to state that these children were the result of a decision NOT to use contraceptives taken by parents who themselves were born in the early 1920s. Better still, we may take the act of marriage as the start of the decision to reproduce, given the Irish pattern of relatively late marriage and the start of reproduction immediately after marriage. These assumptions would not be valid in other countries, or in later periods.

The decision to leave full-time education and to enter the labour market was taken when the median group of potential workers were 15 years old, in the mid-1960s. (This latter decision, also connected to the present supply equation with a very long lag, can even more easily be accepted as a conscious act.)

These time lags are crucial. All the material times (the formative years of the parent generation, and the years when reproduction and educational decisions were taken) relate almost entirely to 'good' years, i.e. periods when employment prospects, in Northern Ireland and in Great Britain, and in the US and other destinations of emigrants, were relatively good. The parents experienced both the post-1933 upswing in labour demand, and the long economic boom which characterized the post-war reconstruction period: an upswing in labour-intensive production occurring just when the very small birth cohorts of the 1930s formed the indigenous annual source of additional labour supply. So short of labour was Britain that there was active recruitment, not only in Ireland, as there had always been, but in the Caribbean and Asia—a boom which lasted well into the 1960s. (In fact, after 1961 the supply of New Commonwealth and Pakistan labour was almost cut off, leaving Ireland as the main source for additional labour.)

The 1960s were the high point of these expectations: not only was there little unemployment in Northern Ireland, but the British construction industry in particular was flourishing, and so were other industries in which Irish labour has always been conspicuous, especially in the public sector. The British GDP was growing at historically very high rates in the 1960s—the ill-fated 'National Plan' of 1965 was based on the assumption of a 5% annual growth rate in the GDP.

These figures are not meant to suggest that high employment chances, high wages, and high mobility are a direct cause of the high fertility rates of the 1960s, which were fairly universal in the Western world at that time. (The rise had begun about 1955.) All attempts to provide rigorous statistical proofs for a one-way causal and positive relationship between real incomes and fertility have failed (Davis, Bernstam, and Ricardo-Campbell, 1986). Nevertheless it is right to put the matter into the form of a conditional one-way relationship: IF parents made conscious decisions about setting up households and having children, they were doing so in the expectation of fast continued growth in labour demand. (This implies also that if and where any section of the population was in the habit of applying restraints to their fertility only if the economic outlook was poor, they had no reason to do so in the 1960s.)

Decisions about education, in so far as they were conscious

and rational, must also have been taken in the light of prevailing labour market conditions. The change to 'post-industrial' economies, as well as accelerated automation of production and services, the boom in computers and robots, did not begin until the mid-1970s, at least in the British Isles. The characteristic shortage of labour from the end of the war until the mid-1970s was in the unskilled and semi-skilled sector. There was the building boom (which reached its peak in 1968) both in housing and other forms of construction. Other types of labour-intensive production all reached their peaks in the 1960s.

## THE EFFECT OF PAST MIGRATION PATTERNS

Net migration out of Northern Ireland had caused a reduction of the population in every decade from 1841–51 until 1891–1901. After that, net out-migration was roughly equal to natural increase until the decade 1921–31. Then emigration again exceeded the surplus of births over deaths. In other words, the generation which provided the parents of today's labour force was born into the perception that there were opportunities for work at home as well as in other countries; the domestic labour supply was kept stable, and was roughly in balance with the demand.

In the war years, and the years immediately preceding and following the war, net migration was considerably reduced, and some time during the twenty years between 1931 and 1951, a turning-point seems to have been reached: the population began to grow again as natural increase once more exceeded out-migration by a considerable margin. Between 1951 and 1961, net emigration rose considerably, and though it never again reached a point where it surpassed the excess of births over deaths, it was large enough to take nearly 10,000 people annually out of the labour market. Between 1961 and 1971, this annual net outflow was reduced to below 7,000 persons, but it was still enough to keep the labour market roughly in balance. (These movements are complementary to those which occurred in the Republic of Ireland, where even more remarkable fluctuations can be observed. These can be directly related to labour demand changes in the Republic and in Great Britain (OPCS, 1980).)

## MIGRATION, NATURAL INCREASE, AND UNEMPLOYMENT

Unemployment in Northern Ireland reached a low of 30,000 (22,000 males) in 1965–6, constituting a rate of 6%. Admittedly that was still twice as high as that observed in the next highest UK region (Wales). This unemployment would correspond to *half* the UK rate at the peak of unemployment in 1986 (after about one-fifth of all formerly counted unemployed persons had been removed from the statistics through various changes in the method of counting). It was a lower rate than that recorded in any UK region in 1987.

The factors affecting the composition of the net out-turn figures for the Northern Irish labour market, convenient for the present analysis, are clear: natural increase is affected both by births and by deaths; total growth, however, in any one year depends on the short-term adjustment by migration. This must be so because there is far more deliberate individual short-term control over migration than over reproduction, let alone deaths. Again this statement applies to a region within one country, not whole countries: internationally, migration is severely affected not only by differences in economic opportunities, but by severe legal restrictions. The scope for adjustment also depends on the relative size of the region under consideration, and the rest of the national labour market.

In 1967, the natural increase rate was at its highest because of the coincidence of the lowest recorded number of deaths in the history of the country with the highest number of births since the 1870s. The natural increase was 19,000; emigration took 7,000 people away, most of whom were adults. Yet unemployment remained low. To be sure, this coincidence does not dispose of the demographic problem entirely: those who died in 1967 were not removed from the current labour supply in that year, nor did those who were born then constitute an addition to the supply. The figures are adduced here mainly to illustrate the long-term lead and lag effects of births and deaths in relation to the labour market.

Summing up, we can say that in the mid-1960s unemployment in Northern Ireland was very low indeed. By the standards of 1988, we would say 6% equalled 'full employment', after taking out of the total the inevitable 'frictional unemployment', and those whom the Department of Employment deems to be ineligible for benefits and unavailable for work. At the same time, migration was actually lower than it had been in the previous decade, despite the boom in the rest of the UK.

We need reminding of this combination of factors, partly because of the general theories which have been made current by Richard Easterlin, notably in *Birth and Fortune* (1980). Without going into his detailed arguments here, we can summarize his thesis by saying that the life chances of children born in any one period depend at least in part on the size of that cohort, and in part on the demand for labour when they reach working age. The size of the cohort is also determined by the expectations their parents (especially their fathers) had when they reached reproductive age. The Easterlin hypothesis effectively provides a two-generation lagged effect in determining current labour supply. We can use this theory at least as a framework for a summary of a working hypothesis we can now state: the very large number of entrants into the Northern Ireland labour force in the 1980s is to a large extent due to a mood of optimism amongst their parents in the 1960s; a mood influenced by very low unemployment, and also in part by the ample opportunities for out-migration. The size of the entrant cohort may also reflect the experiences parents had, much further back in time, when they were entering the labour market, in the 1930s or 1940s.

## TWENTY YEARS ON

Since that time everything has changed, except the number of children who survive to become part of the labour force. Emigration has fallen considerably: it fell by half between 1966 and 1986. This means that the size of the cohorts entering the labour market has not been much reduced by migration: of the

0–4 age group in 1971 (including the years of the largest birth cohorts), 92% were estimated still to be resident in 1986. Of the missing 8%, 2% would have died (applying current life tables). This suggests that less than 5,000 boys were removed from the mid-eighties entrant labour force cohort by migration.

The number of deaths has risen since the low point of 1967: in 1985, when births were at their lowest point, the natural increase was only 11.7 per thousand, compared to 17.6 in 1967–8. With the fall of net emigration to under 4,000, this means the rate of population increase is now about 7,000, i.e. it is much lower than it was in 1967–8.

It will be seen that even if reproduction (fertility) were a totally voluntary act, it could not by itself determine the rate of increase of the labour force. Deaths are a function of mortality (which has fallen as much in Northern Ireland as in the UK as a whole); and they are also a function of what has happened to the original cohort: this, in turn, is determined by fertility at the beginning of the twentieth century, which was still very high, but also by infant and child mortality in the years before the First World War.

The most important determinant of the size of the cohort which will die in the current period, however, is the size and composition of migration streams since the time of its birth. Since between 1901 and 1931 net out-migration nearly equalled natural increase, it may be said that there are now 'fewer people at risk of dying', i.e. a factor unrelated to past or current mortality. Since the average annual number of births from 1926 to 1937 was the lowest ever recorded (the years are not arbitrarily chosen but are determined by the vagaries of the Northern Irish census dates), it may be assumed that the number of deaths will fall between now and the end of the century: the 'hollow' in the Northern Irish population pyramid is clearly visible in each successive census enumeration.

This imbalance should also be considered when looking at these age pyramids as a whole. When we compare the age structure of Northern Ireland with that of Great Britain, in percentage terms for each age group, it seems striking that the older age groups are underrepresented, and the younger ones overrepresented (Eversley, 1989). This difference is, however, as much due to the loss from the older groups by migration over most of this century, as it is to the greater fertility of the most recent periods.

It is clear, therefore, that past events have a much stronger effect on the size of the potential working population than any current changes in such parameters as economic activity rates. This is generally true, but even more so in Northern Ireland, where the rate of emigration, especially of the adult population, has always been much higher than for any other part of the UK (except possibly the Scottish Highlands during the Clearances). Although, after 1900, decadal losses of 10% of the population did not occur again, they still amounted to 6.7% between 1951 and 1961. This not only affects the current unemployment rate, but also fertility, since it is the most fertile age groups which are liable to emigrate. If, as in Ireland, there is also a severe sex imbalance in emigration rates, the total effect on reproduction is even greater than total losses of adult population suggest.

## THE EFFECT OF THE LAG

Not only do most people not make wholly rational choices about reproduction in the early years of marriage, but there is also no way they can foresee the future. If asked about intended family size, they make necessarily rather uninformed guesses about the future and their own position in society (Compton and Coward, 1987). Thus the fact that the largest cohorts, relative to the existing employed labour force, entered the labour market in the early 1980s, and that emigration fell sharply at the same time, cannot be blamed on the parents of the 1960s. We say 'existing employed labour force' because there is not a simple relationship between school-leavers and people reaching retiring age.

Rather, it is a question of how many persons (males in particular) in the older age groups are still at work, and therefore can theoretically vacate a workplace by retiring. As we have shown elsewhere (Eversley, 1989), the size of the entry (15–24) group was particularly large just when the pre-retirement age group (55–64) actually working was particularly small, partly because so many of that older group had emigrated to Great Britain in mid-life, and partly because very few of them still had jobs when the size of the entrant group began to rise.

The proportion of entrants to leavers can be calculated in census years: in the rest of the UK, in 1981, this varied between 1.5:1 in the most prosperous regions to 2:1 in the least favoured ones. In Northern Ireland, the proportion had deteriorated from 1.66:1 in 1971 to 2.0:1 in 1981. (These figures include all males, whether working or not.) If we break down this proportion within sub-regions and within individual districts, distinguish between Catholics and 'others', and allow for the older men who were no longer working, we find that the figure rises to over 6:1 in some places (it was well under 2:1 for non-Catholics in many areas).

The instinctive reaction of the commentators has been to blame these severe imbalances on the large number of Catholic children. This, as we have shown, is far from being correct: a great many other factors play a role. If, for example, there were nearly seven male Catholic work force entrants in some western districts for every one person still in employment in the pre-retirement group, this is in part due to high unemployment in the older group; in part to past emigration patterns; in part also to differentially greater mortality at the time when these older men were children.

To add to the pressures, whereas the pre-retirement age group is almost exclusively male, the entrants also comprise women, to an increasing extent. To be sure, there should also be women leaving the labour market to balance the entrants. But practically all women now join the labour market, and economic activity rates for older women are particularly low in Northern Ireland. The older women are almost exclusively in low-paid, part-time, unskilled jobs; the new entrants include a higher proportion of girls with skills, many of which are not gender-specific (i.e. they can take jobs previously held by men), and thus the pressure is still further intensified.

How much is due to each factor? That cannot be precisely calculated, particularly at the level of the local labour market.

However, we do know that differentially far more Catholics emigrated during the last fifty years (Barritt and Carter, 1962; Compton, 1986), but how many of these left each district, and at what age, is impossible to calculate. Attempts have been made to model the past and present population by district, with a view to dividing the changes into the effects of natural increase and of migration, but no convincing results have been produced (Eversley and Herr, 1985). Migrants from any one area may have gone to another part of Northern Ireland, the Irish Republic, Great Britain, or overseas. Relatively precise information for some of these categories is available only for census years. We can thus only enumerate contributory factors, but not break them down in detail.

RELEVANCE TO CURRENT UNEMPLOYMENT

Given a largely segregated labour market, where Catholics have tended to take up jobs vacated by other Catholics, and Protestants to take Protestant jobs, past patterns tend to perpetuate themselves. Thus if a high proportion of older Catholics are not working, in any one area, there are fewer places open to young Catholic labour force entrants. Past patterns of discrimination and disadvantage will perpetuate themselves, unless legislation against discrimination and the promotion of equality of opportunity is more effectively enforced. If there is low geographical mobility between areas (whether at the level of neighbourhoods, or of districts and even sub-regions), this constitutes a further factor producing rigidity.

This immobility is in itself partly a result of discrimination (both actual and perceived), and partly of a feeling of physical insecurity: people will not cross certain boundaries in search of work. How much this is due to a documented history of violence, and how much to somewhat irrational fears, is immaterial: under the conditions of Northern Ireland society, we have to take this fear as being a real factor. Thus the local as well as the national demographic anomalies of the labour market perpetuate themselves.

## DIFFERENTIAL FERTILITY

In the older textbooks of demography, differential fertility figures as an important phenomenon. It has therefore been convenient to point to one of the best-documented Western fertility differentials, that between Catholics and Protestants, as the principal cause of Catholic unemployment in Northern Ireland. This view has survived to this day. There is, however, a somewhat parochial perspective in many writings on the situation in Northern Ireland. Although such a differential is well attested for past generations, in many countries, it has not usually been cited as a cause of excess unemployment amongst the Catholic populations. To be sure, there have always been countries where Catholics have tended to live in rural areas, often as peasants, whereas Protestants were urban, working in trade and manufacturing industry. Peasants may have lower incomes; but it is not suggested by writers on this differential that the peasants are poor because they have more children than the Protestants.

In the decades since the 'Population Scare' really got into its stride (after about 1955), it has been fashionable to blame poverty on the fertility of the less-developed countries. This convenient alibi has come to be less frequently mentioned in the last few years as the essential fallacy behind the argument has appeared.

We cannot enter into the details of this debate here; but it has been pointed out that high fertility and low rates of economic growth do not correlate very well either for our time, or, *a fortiori*, for the years of most rapid economic growth in Western countries in the past. These were periods when fertility was very similar to what it is in the less-developed countries today, and they were the years of greatest economic growth. At other times very low fertility coincided with economic stagnation or even decline. For this reason, excessive fertility as the cause of poverty figures much less in current literature, except for a small group of anti-population propagandists mainly in North America, and those Northern Ireland authors who cling to the older view.

## THE DEMOGRAPHIC TRANSITIONS AND
## DIFFERENTIAL FERTILITY

We ·do know that Catholic fertility was, and is, higher in Northern Ireland, than it is for the 'others'. The 'excess' fertility has varied over time. It can be shown that the process known as the demographic transition has a special variant in Northern Ireland.

There was, in the nineteenth century, a first demographic transition when, by means not wholly understood even now, first mortality and then, after a lag, fertility dropped in most countries, starting with France, and reaching most other (industrializing) countries by the early twentieth century (Peterson, 1975). The second such transition took place much later: from a new high reached just after the Second World War (still low compared to levels prevailing at the beginning of the century) to a new low point about thirty years later. (The course of events is slightly different in the USA, but this does not affect the sequence.)

The fall continues in certain countries like Germany, but there has been some recovery since the lowest points were reached in the late seventies. Two countries only, Ireland and Poland, have failed to participate fully in this second transition. We advisedly say 'Ireland' because this statement applies both to the Republic and to Northern Ireland (Sexton and Dillon, 1984; Coward, 1980). It is easy to come to the quick conclusion that these are two fervently Catholic countries unlike Italy or France where the church does not have the same hold on the inhabitants. Such a facile conclusion is as wrong as an equally facile alternative: that these are two countries with intensely nationalistic feelings. A moment's reflection on other countries showing similar patterns of church influence, and of nationalism, proves the futility of such generalizations.

One fact, however, does stand out: in most countries where there are mixed populations, the majority usually adopted the second transition before most of the minority groups. There are exceptions: thus the Jews, wherever they lived in modern times, controlled their fertility before the Christian host communities; the Chinese in parts of East Asia adopted contraception before the majority ethnic and religious groups. Why this should be so is beyond the scope of the chapter.

Certainly, in Northern Ireland, the changing relative fertility of Protestants and Catholics points to the conclusion that the Protestants were the first to engage in this second transition, leading to a widening in the fertility gap (Coward, 1980; Compton, 1986). The Catholics followed after an interval, starting with the post-war high of fertility for both groups (about 1964) and proceeding to the latest available information, which shows the gap to have narrowed considerably (Compton and Coward, 1989).

Explanations vary: some of the differences *may* be of purely religious origin, i.e. they may be due to the observance, by Irish Catholics in Ireland, of the church's prohibition of contraception (and of course abortion). Some of it may be economic-structural: in all countries, populations with high economic activity rates for women tend to have lower fertility; this applies also to subgroups. In general, in the West, in the most recent period, the fertility of the higher socio-economic groups has still been lower than that prevailing at the bottom end of the scale.

Rural populations in many countries still have higher fertility than urban populations (except for ethnic minorities). These are well-known persistent differentials. One additional factor has recently been observed in Great Britain: areas of high unemployment have had significantly higher fertility rates than those of low unemployment (CSO, 1988). (Not because 'the unemployed have nothing better to do', or because 'to have more children is the only way a family on supplementary benefit can raise its income', but for complex reasons which are not yet fully understood. The purpose of this statement is that the vulgar generalizations are easily disproved; to prove the truth of an alternative explanation is much less easy. The absence of provable explanations is, however, no reason for accepting simplistic inferences.

What does this mean for Northern Ireland? The conventional wisdom of some of the commentators has it that socio-economic explanations have little significance; the difference is said to be 'cultural'. Demographers in fact do not normally use any such category. The conclusion must be that all the factors mentioned *may* play a role: income levels, social benefit support systems, the attitude of the church, availability of contraceptives, and the ease of obtaining abortions (in England). These cannot be quantified.

Nor can other cognate matters be given paramount explanatory significance. One such subject is the persistence of many small family businesses in Northern Ireland, compared to the rest of the UK: farms, shops, artisan enterprises. This applies to Protestants and Catholics equally (Eversley, 1989); consequently, one might surmise, both groups could have higher fertility than comparable populations elsewhere, where self-employment is rarer. There is still some truth in this. (The facile comparison with Poland again surfaces, and can be shown to be wrong in a largely collectivized economy.) There is more work for children in the family economic unit in Ireland than there is in England. British commentators analysing the persistence of high fertility in certain communities of Asian origin like to point to the prevalance of family-run businesses, especially in retailing, as a supposed reason for not adopting family limitation. The error lies not in mentioning the possibility of some of these factors having some significance at one point in time, or in one locality, but in choosing one to take the entire burden of explanation, perhaps because it is politically expedient to do so.

## FAMILY SIZE AND ACCESS TO THE LABOUR MARKET

We can accept that the size of Roman Catholic families (or households—the two are not interchangeable) in Northern Ireland is still significantly larger than it is amongst Protestants. The actual size of the differential at any one moment has in it an element of time distortion: thus, families (or rather households) in which all the children date back to the years of high European fertility (the late sixties and early seventies) may have one more child than Protestant families; families whose entire reproductive span falls into the years after 1977 (and whose fertility may not therefore be completed) may have a differential of only 0.5 children (Compton and Coward,

1989). In the next decade, the differential may still be the same proportionately, though at much lower levels.


## AN 'EXPLANATION' OF HIGH UNEMPLOYMENT?

First, a *reductio ad absurdum*. Whatever questions applicants for jobs may be asked, it would be bizarre if they included this one: how many brothers/sisters have you at home? If it were to be adduced as a reason for not giving the applicant the post that he/she is burdened with too many siblings, family size would play a role. This is unlikely to be the case. In theory, a child from a large family should have the same chance of a job as one from a small family. So family size *per se* cannot be a reason for excess unemployment.

We move on to firmer ground if we ask whether family size is related to the location of the family. In modern labour market economics, it is sometimes claimed that those with fewer children have a better chance to move (more housing choices, more likelihood of benefiting from a second earner in the household (Peterson, 1975)). In Northern Ireland, in the years before the present Housing Executive took over the management of the stock and the allocation of new houses, discrimination was widely alleged (Eversley, 1989). But even then, the allegation was not that the housing authorities gave preference to smaller families: simply that they did not allocate houses to Catholics at all, in areas where there was work. (We have to remember that we refer to those years when Northern Ireland still had virtually full employment.)

Since then, the allocation of housing has been much fairer, and complaints on that score are now few. The evidence both of the 1981 census and of the Continuous Household Survey is that Catholics do indeed live at higher densities than Protestants (with about one more child for each household with dependent children). This is not a bar, however, to the occupancy of a house in a Catholic housing estate. It might, indeed, be more pleasant to the prejudiced eye of an English middle-class observer, if families could have one bedroom per child, and a spare for visiting relations; but this may not be the preferred allocation of spending patterns by any one community, as we know from multi-racial societies in all countries. There are differences, of course, by social class, or socio-economic group, both for Protestants and for Catholics; but they have no explanatory significance when looking at unemployment.

In the literature, there is a good deal of reference to the general immobility of Roman Catholics 'trapped' in areas of low incomes and employment opportunities. Without trying to pass judgement on this view, it certainly is much less likely now that larger families find it harder to move within the country. Such an assertion would be based on the idea that larger families are adequately housed in larger dwellings in the peripheral areas, and more cramped in the Belfast core region. That is not the case, on the evidence of the 1981 census, or of the Continuous Household Survey so far as it distinguishes between localities. So there must be other reasons for their inability, or unwillingness, to move.

It can be shown from the evidence of British census reports since the war that migrants from Ireland included a higher than expected proportion of single adults, that is, compared to the population of Ireland as a whole. Most Irish workers did not bring their families, if any, with them. We cannot say whether this was because they chose not to marry until they had earned an adequate income for some years; or because of the high price of accommodation, especially in London, for families with children; or because they remained, essentially, seasonal migrants—as British workers, as well as Irish ones, frequently had been in the nineteenth century. There is no evidence from this aspect of migration statistics that family size is a bar to internal mobility within Northern Ireland. There is, however, evidence that family size is an obstacle to movement from any region, Irish or British, into the southern English areas of fastest economic growth, lowest unemployment, and highest wages. This observation is not specific to religion or race.

We now turn briefly to other alleged obstacles to employment, like lack of education and training, that again cannot be linked to demographic structure. That Catholic children found it hard to obtain relevant education and training in some of the peripheral areas in the 1950s and 60s is not doubted. That was so, however, because of the inadequacy of the educational institutions, not the size of families. Once again, it is the past that overshadows the present: the cohorts leaving school or further education in 1988 may be as well qualified in Catholic as in Protestant areas, but because this was not the case twenty or thirty years previously, the difficulty persists (see Chapter 4). No Catholic child would have had a better education in these earlier years if it had been an only child.

One last aspect of differential fertility can be discussed. The literature on the causes of fertility decline is still far from unanimous about its causes. Fertility has declined for countries with totally different political systems, degrees of economic success and real income per head, and levels and trends of real wages (Eversley and Koellman, 1982). In almost all societies, however, there has been a tendency for clerical workers to have fewer children than either manual workers or managerial/ professional workers. Cause and effect are difficult to separate. There are complex and elegant theories about status, the position of women, and the requirements of a domestic environment (Ermisch, 1982). It is a vulgar misinterpretation, however, to say that this amounts to the reduction of fertility becoming a pre-condition for upward social mobility. In the case of Ireland, a country where marriage took place very much later, for at least a century, than almost anywhere else in the world, and where these careful bachelors and spinsters had as their reward for their Malthusian prudence a lifetime of abject poverty, the over-simplified theory is particularly inappropriate (Kennedy, 1973).

However, we may well accept the more moderate view that larger prospects of future professional advancement may be associated with more widespread adoption of birth control. (Family size is indeed lower in those Belfast suburbs where the civil servants and professional people congregate, and this applies to Catholics as well as Protestants.) If that is so, then we can turn the argument on its head and say that since Catholics were consistently denied promotion to the better positions in Northern Irish society, they had no particular motivation to adopt birth control, apart from the fact that it was against their conscience.

## CONCLUSION

The observer of the Northern Irish scene is surprised to discover that arguments which have been ruled out of court in the US (against Hispanics and indigenous blacks), and in Britain (against West Indians twenty years ago, and Asians more recently), which effectively blame the low socio-economic position of the minorities on their fertility, still flourish in Northern Ireland. The facts certainly do not bear out the allegation. We have shown that the admittedly still rather large cohort of children (Catholic and Protestant) who have survived to be adults in the present labour market, and who suffer very high unemployment, were born during a period when numbers presented no special problem. Relatively good employment chances in the Province, and, for some time in the 1970s in the Republic as well, plus a steady demand for Irish labour in the rest of Britain, made fertility irrelevant. Admittedly, there always were relatively far more Catholics than Protestants unemployed, and in lower-paid positions (Hepburn, 1982), but at a time of generally low unemployment and with a steady rise in both real earnings and social benefits, and services, the enclaves of high unemployment in some western districts were of mainly local significance.

It should be noted, however, to deal with yet another facile generalization, that the start of the new wave of acute unrest, an increase in both nationalist aspirations and of loyalist opposition, dates from a year when prospects in the labour market were arguably at their best, both in Ireland and in Great Britain. Although the political problem certainly has its economic roots, employment problems did not cause the new wave of strife. (The same observation can be made about the student revolution of 1968: that too had no visible economic causes.)

The children who had the misfortune to be born in Northern Ireland in the sixties and early seventies, whether only children or part of a family of four or five, Catholic or Protestant, have arrived on the labour market at a time when everything militates against them. This is certainly a large element in the present continuation of the struggle, and affects both the majority and the minority: they fear for their future. The causes are manifold, but the size of the Catholic family in the past is not one of them. There is no evidence at all that if Catholics had had no more children, in the past, than Protestants, their unemployment rate would now be lower. All the evidence is the other way.

There remains only one demographic aspect which cannot be totally ruled out: if Northern Ireland had fewer entrants into the labour market in the 1980s, Catholic and Protestant, it is just possible that fewer of them would now be unemployed. The same has been said of the size of the (very largely Protestant and white) cohort in Great Britain at the same time; and the reduction in birth cohort size in the seventies is cited as one of the reasons why employment prospects are likely to be better in the nineties. Under 'demographic causes' we must, however, include also the relative sizes of birth cohorts in much earlier times, and especially, in Northern Ireland, the composition of migration streams.

In sum, demographic causes do not explain high unemployment in Northern Ireland; they throw even less light on the proportionately much higher unemployment in the Roman Catholic population.

# 4

# Educational Qualifications and the Labour Market

*Robert D. Osborne, Robert J. Cormack, and Anthony M. Gallagher*

## INTRODUCTION

In the Fair Employment Agency's first annual report it was indicated that the Agency proposed to develop a research programme that would seek 'to begin to identify and explore the various factors whose interaction generates the differing patterns in employment and occupations' (FEA, 1978). One of those factors soon identified was educational qualifications. Two research papers followed which examined the extent of educational differences between the two communities and the significance, if any, of these differences for employment (Osborne and Murray, 1978; Osborne, 1985). Interestingly, educational research in Northern Ireland has tended to concentrate either on traditional educational concerns such as selection, management, and teaching methods (see, for example, the work of the Northern Ireland Council for Educational Research) or on the implications for intercommunal relations of having a segregated school system (see Gallagher (1989) for a review of this latter research). This chapter seeks to outline the evidence now available on the nature and extent of differences in educational qualifications between the two communities, to assess the reasons for these differences, and to consider the implications for employment. The second area of analysis extends the consideration of the education system to higher education. Drawing on major surveys, participation rates since the early 1970s of the two communities in higher education are considered and the early careers of graduates are analysed.

## THE STRUCTURE OF EDUCATION IN
## NORTHERN IRELAND

At a general level, the structure of the education system in Northern Ireland is relatively straightforward. The system consists of primary schools attended by pupils from the age of 5 to 11. At the age of 11 pupils are separated into grammar and secondary intermediate schools principally by means of the Transfer Procedure (see Wilson, 1987). (In the Craigavon area, a comprehensive-style junior and senior high school system exists.) A substantial proportion of grammar schools also take fee-paying pupils. At all stages of the education system there is a separation of pupils on the basis of religious affiliation. The Catholic authorities negotiated the provision of separate schools with the Northern Ireland government in the early 1930s. State funding of these schools increased over the years to the current position where all 'maintained' (i.e. Catholic primary and secondary intermediate schools) receive all recurrent costs and 85% of capital costs. Arrangements were also made between the Protestant churches and the Northern Ireland government for the transfer of their schools during the 1920s and 1930s. As a result these state or 'controlled' schools (primary, secondary, and some grammar) are now administered by public authorities, although the Protestant churches retain transferor's rights in terms of membership of school governing bodies etc. and are fully funded by the state. In addition, there are a large number of voluntary grammar schools: all Catholic grammar schools and some of the grammar schools in the Protestant sector. These schools are fully funded by the state for revenue costs and receive 85% of capital costs. Although there is not a Protestant school system in the same way as there is a Catholic system, the non-Catholic schools are *de facto* Protestant in terms of the pupils attending them and teaching staff. There is no recent research assessing the extent to which pupils 'cross over' to the 'other' side. It is believed, however, that the scale of this is quite small, being confined to Catholics attending some Protestant grammar schools particularly in the Belfast area and some Protestant primary schools where demographic changes and population movements have led to Catholics having no local Catholic school within reasonable travelling distance.

The institutions of higher education in Northern Ireland are 'integrated' in the sense that the two universities are not owned or controlled by the churches and students from both traditions attend the two universities. The universities' patterns of employment are considered in Chapter 2. Undergraduate teacher training, which is not considered here, is conducted partially through segregated teacher training colleges and also through the universities.

## ACCESS TO GRAMMAR SCHOOLS

One of the major issues that has emerged in recent years is the extent to which access to grammar places is similar in the two systems. Recent research has suggested that when the number of primary pupils at the transfer stage is related to grammar places in the same system, the Protestant system has a significantly higher ratio of places—36% to the Catholic system's 28% (Osborne *et al.*, 1989). Part of this difference is related to the higher incidence of fee-paying places in the Protestant system. There is, however, a marked difference in the geographical availability of grammar places in the two systems, as Figure 4.1 reveals.

If we take the 30% cut-off point as approximating the average proportion of pupils transferring from primary schools and awarded a free grammar school place by the Department of Education for Northern Ireland (DENI), it is quite clear that many more areas of Northern Ireland show a much lower availability of Catholic grammar school places in relation to primary pupils than is the case for Protestants. This factor, combined with the higher ratio of fee-paying places in Protestant grammar schools, imparts a strong *structural* explanation for lower proportions of Catholics in grammar schools. However, Livingstone (1987) has recently analysed data which reveal that in the transfer procedure in 1984–5 approximately 10% of Catholic children who had achieved a grade entitling them to a free grammar place subsequently transferred to a secondary school. This suggests a *choice* explanation for lower

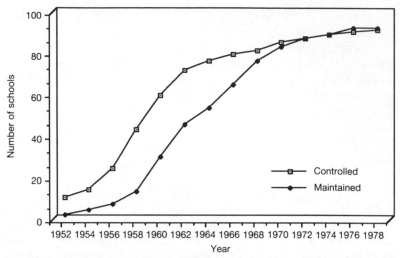

F IG. 4.1. Number of controlled and maintained schools: Northern Ireland
1952–1978

Catholic representation in grammar schools. In other words, some parents choose not to place their child in a grammar school. However, Livingstone was unable to pin-point what proportion of the 10% were unable to select a grammar school for reasons associated with geographical access. Some evidence suggests that in West Belfast choice is a factor in that three secondary schools are functioning as comprehensive or all-ability schools since a substantial number of pupils are attending them despite having been awarded a free grammar school place (having passed the eleven-plus). However, this is likely to be a phenomenon specific to this area (Armstrong, 1989). On the significance of 'choice versus structure' throughout the Province, further research is under way by the authors to resolve this issue. The different proportions of Catholics and Protestants has important consequences for the output of schools, as we discuss below.

## QUALIFICATIONS AND JOBS

The expansion of certification in the education system during the 1950s, 1960s, and 1970s represented an attempt by the education system to offer pupils and parents the incentive of externally validated examinations which would have a clear and recognized value in the labour market. It was argued that access to jobs would increasingly become dependent on the possession of these qualifications—the so-called 'tightening bond' thesis (Marshall, 1963). Researchers monitoring the transition of school leavers into the labour market during the 1970s argued that there was clear evidence of the tightening bond so that even quite marginally qualified school-leavers seemed to fare better than the unqualified (Gray *et al.*, 1983). Raffe (1984) reported the strong connection between possession of Scottish O grades and entry to employment. Raffe and Willms (1989) also suggest that unemployment levels in local labour markets can have a significant influence on 'staying on' rates but that different groups of young people respond differently depending on the structure of employment.

However, researchers taking a broader view of social mobility have also drawn attention to the 'non-meritocratic' factors associated with occupational access. Thus, for example, Heath and Ridge (1983), using the 1970s social mobility data for England and Wales, showed that service-class boys attending grammar schools from 1944 had actually increased rather than decreased their advantages over working-class boys. Post-war economic expansion made it easier for the 'service class' to strengthen its hold despite the expansion of credentialism as a means for the selection and promotion of people in the labour market. Similarly, Payne and Ford (1979), drawing upon Scottish data, suggest that, while the possession of formal qualifications is often a sufficient condition for access to non-manual occupations regardless of class background, it is not a necessary condition; many people obtain positions, particularly in industry and commerce, even in the absence of qualifications. To these writers, this provides a new focus of research—the role of institutions as recruiters, selecters, and promoters mediating between educational qualifications and access to jobs.

Researchers who have looked at the activities of employers have tended to concentrate on the recruitment of school leavers, with most of the research being undertaken at a time of high unemployment. They found a mixed picture as far as the treatment of school-based qualifications is concerned. For jobs with a technical content, specific qualifications or experience are often sought, while, in other contexts, employers have used academic qualifications as a means of screening large numbers of applicants for a small number of jobs. Academic inflation, or 'creeping credentialism', has tended to involve employers asking for ever higher levels of qualification, (Ashton and Maguire, 1980). These studies date from the period of large-scale school-leaver unemployment and it may be that the rapid decline in the size of the school-leaver cohorts of the 1990s in the UK will alter the ways in which employers make use of educational qualifications in recruitment.

## OUTPUT FROM SCHOOLS: QUALIFICATIONS OF LEAVERS

The main source of data for examining the qualifications of school-leavers is the School Leavers Survey (SLS), conducted annually by the Department of Education for Northern Ireland. Unlike the similar exercise undertaken in England and Wales, the SLS in Northern Ireland gathers data for all school-leavers rather than on a sample basis. Data are available for a number of years between 1975 and 1987 (Table 4.1). Attainment levels in the two systems have improved in the period under review but the scale of difference between the two systems has not greatly changed. Overall, a higher proportion of Protestants leave school with A levels and O levels whereas a higher proportion of Catholics leave school without qualifications. In 1987 it would have required a reduction of 19.7% in the numbers of Catholics leaving school without qualifications for the Protestant figure to be matched.

The differences in the qualifications of leavers are influenced by the different proportions of Protestants and Catholics in grammar schools, as discussed above. In Table 4.2 are shown the qualifications of leavers broken down by school type for 1987. The differential pattern of attainment evident at the aggregate level can be seen to be attributable, at least in part, to the larger proportion of Protestants in grammar schools.

There are also significant differences between the attainment

TABLE 4.1. *Qualifications of school leavers by religion, 1975 and 1987 (%)*

|  | 1975 | | 1987 | |
|  | Protestant | Catholic | Protestant | Catholic |
|---|---|---|---|---|
| 3 or more A levels | 11.8 | 7.5 | 14.8 | 12.0 |
| 2 A levels | 5.5 | 5.6 | 6.0 | 6.7 |
| 1 A level | 3.6 | 3.3 | 3.8 | 3.6 |
| 5+ O levels/Grade 1 CSE | 9.3 | 7.8 | 12.8 | 10.8 |
| 1–4 O levels/Grade 1 CSE | 18.3 | 17.3 | 23.7 | 23.3 |
| CSE Grades 2–5 | 10.9 | 12.1 | 20.7 | 21.1 |
| No qualifications | 40.6 | 46.4 | 18.2 | 22.5 |
| TOTAL | 100 | 100 | 100 | 100 |

*Note*: Data for grades 2–5 CSE also include grades D–E O level in 1987. Qualifications other than GCE/CSE are not included.

*Sources*: Data for 1975 adapted from Osborne and Murray, 1978, and for 1987 from Osborne *et al.*, 1989.

TABLE 4.2. *Qualifications of school leavers by school type and religion,*
*1986–7* (%)

|  | Catholic | | Protestant | |
|---|---|---|---|---|
|  | Grammar | Secondary | Grammar | Secondary |
| 3 + A levels | 40.1 | 1.7 | 41.9 | 1.0 |
| 2 A levels | 20.2 | 1.8 | 16.6 | 0.7 |
| 1 A level | 9.2 | 1.5 | 9.8 | 0.8 |
| 5 + O levels/ Grade 1 CSE | 12.5 | 10.0 | 15.5 | 11.4 |
| 1–4 O levels/ Grade 1 CSE | 14.4 | 26.5 | 14.1 | 28.0 |
| CSE 2–5/Low O levels | 1.8 | 28.2 | 1.1 | 30.6 |
| No qualifications | 1.8 | 30.2 | 0.9 | 27.5 |
| TOTAL | 100 | 100 | 100 | 100 |

*Source*: Osborne *et al.*, 1989.

levels of Protestants and Catholics when gender is taken into account. The data are shown in Table 4.3 for 1987. For both Protestants and Catholics, the attainment levels of girls are better than those of their male co-religionists. The position of Catholic males is especially worthy of comment. The proportion of Catholic males leaving school with A level qualifications is broadly the same as the proportion of Protestant males, the position of Catholic males is markedly worse than their Protestant counterparts in other categories. Thus, while Catholic males represented 23.6% of all school leavers in 1987, they represented 34.1% of those without qualifications.

Geography is the final dimension which imparts variation in the patterns of school leavers' qualifications. In Table 4.4 are shown the proportions of those qualified with A levels and those leaving with no qualifications. For four of the Board areas similar proportions of Catholics and Protestants leave school with A levels, but, in the Belfast area, Protestants are markedly more likely to leave school with A levels (39%) than Catholics (27.7%). This is, in part, a reflection of the concentration of

TABLE 4.3. *Qualifications of school leavers by religion and gender, 1986–7* (%)

| | Catholics | | Protestants | |
|---|---|---|---|---|
| | Males | Females | Males | Females |
| 3 or more A levels | 11.4 | 12.6 | 13.2 | 16.5 |
| 2 A levels | 5.9 | 7.5 | 5.0 | 7.0 |
| 1 A level | 2.8 | 4.3 | 3.2 | 4.5 |
| 5 + O levels/ Grade 1 CSE | 8.1 | 13.6 | 11.1 | 14.5 |
| 1–4 O levels/ Grade 1 CSE | 20.6 | 26.0 | 21.9 | 24.8 |
| CSE 2–5/Low O levels | 21.7 | 20.5 | 21.7 | 19.6 |
| No qualifications | 29.5 | 15.5 | 23.9 | 13.1 |
| TOTAL | 100 | 100 | 100 | 100 |

*Source*: Osborne *et al.*, 1989.

Protestant grammar places in Belfast, which results in pupils being drawn in from outside the city (see Figure 4.1).

The actual grades received by those passing O and A levels in 1979 and 1982 are shown in Table 4.5. As can be seen, Protestants record a higher percentage of passes at Grade A at both examination levels for the years in question. Although the data are not presented here, this attainment difference extends across subject categories. In terms of the overall trends that can be discerned in these patterns, the evidence of convergence during the 1970s seems to have levelled off and the 1980s have seen something of a plateau in the relative levels of achievement between the two systems.

EXPLANATIONS

The difference in attainment levels of children from the two school systems has been identified as largely the result of two factors: the differential provision of grammar schools in the

TABLE 4.4. *Proportions of school leavers with A levels and with no qualifications by religion and Area Board, 1986–7 (%)*

|  | Catholics | | Protestants | |
|  | With A levels | No qualifications | With A levels | No qualifications |
| --- | --- | --- | --- | --- |
| Belfast | 27.6 | 24.7 | 39.0 | 16.9 |
| Western | 20.4 | 25.7 | 21.5 | 20.8 |
| North-Eastern | 20.8 | 18.9 | 18.9 | 20.8 |
| South-Eastern | 19.2 | 20.4 | 21.2 | 15.6 |
| Southern | 22.4 | 26.8 | 25.6 | 19.6 |

*Source*: Osborne *et al.*, 1989.

TABLE 4.5. *Proportions of O and A level passes obtained at grade A by religion, 1979 and 1982* (%)

| Religion | O level | | A level | |
|---|---|---|---|---|
| | 1979 | 1982 | 1979 | 1982 |
| Protestant | 18.5 | 18.9 | 14.6 | 15.2 |
| Catholic | 13.2 | 13.9 | 12.0 | 13.0 |

*Source*: Osborne, 1986.

Catholic system as outlined above and the different social class background of Protestants and Catholics. The Catholic unemployment rate is twice the rate of Protestants, and Catholic males are more likely to be in semi-skilled and unskilled jobs (Osborne and Cormack, 1986, 1987; Chapter 2). The Catholic community as a whole has, on average, larger families, a factor often linked to lower attainment levels, and is more likely to be dependent on social security benefits (Harbison, 1989). These socio-economic differences raise interesting questions about the intake to the education systems. To what extent does 'class' or social background influence attainment in the two school systems? Are Catholic schools able, in some way, to mitigate some of the effects of social background, which elsewhere has consistently proven to be a highly significant determinant of educational attainment (Rogers, 1986)? Some research in Scotland suggests that this may be the case with Catholic schools there (Willms, 1989). Does the larger average family size of Catholics influence attainment levels? These questions require further research to enable the workings of the two systems to be more fully understood.

QUALIFICATIONS: SUBJECT BALANCE

A second aspect of the output of schools relates to the subjects being studied in the two school systems. Educational researchers in Northern Ireland have pointed to different approaches to common subjects, such as history, in the two school systems but have generally failed to assess whether there were any aggregate differences in the balance of subjects taught. It was not until the linking of educational issues to labour market concerns through the fair employment issue that this matter was investigated.

In the assessment of curricular differences two data sources have been used. The first uses examination passes at GCE O and A level gained by individuals in the two school systems. Data are available for the years 1967, 1971, 1975, 1979, and 1982 (although the data for 1967 are less complete than for later years). By assessing qualifications gained in this way we are reflecting our concern with the nature of qualifications gained and their currency in the labour market.

TABLE 4.6. *Subjects of O and A level passes by gender and religion, 1975 and 1982* (%)

|  | Catholics | | | | Protestants | | | |
|  | 1975 | | 1982 | | 1975 | | 1982 | |
|  | M | F | M | F | M | F | M | F |
|---|---|---|---|---|---|---|---|---|
| O level | | | | | | | | |
| English Language and Literature | 21 | 27 | 21 | 25 | 22 | 27 | 20 | 25 |
| Maths | 18 | 11 | 16 | 13 | 20 | 14 | 21 | 14 |
| Sciences | 14 | 8 | 17 | 14 | 22 | 13 | 23 | 16 |
| Languages | 19 | 20 | 13 | 14 | 9 | 16 | 8 | 13 |
| History/Geography | 14 | 12 | 12 | 11 | 16 | 16 | 14 | 14 |
| Arts | 9 | 14 | 11 | 18 | 5 | 7 | 6 | 13 |
| Crafts | 3 | 0 | 4 | 0 | 5 | 0 | 6 | 0 |
| Other | 2 | 8 | 5 | 5 | 6 | 2 | 2 | 4 |
| A level | | | | | | | | |
| English | 11 | 23 | 8 | 18 | 8 | 20 | 7 | 16 |
| Languages | 17 | 24 | 10 | 15 | 6 | 20 | 6 | 14 |
| Humanities | 29 | 21 | 25 | 19 | 23 | 21 | 21 | 23 |
| Mathematics | 14 | 6 | 20 | 8 | 19 | 8 | 21 | 11 |
| Sciences | 23 | 10 | 31 | 19 | 36 | 19 | 37 | 20 |
| Crafts | 1 | 2 | 0 | 3 | 3 | 5 | 2 | 9 |
| Arts | 5 | 12 | 6 | 15 | 4 | 8 | 6 | 5 |
| Other | 0 | 2 | 0 | 1 | 1 | 0 | 0 | 1 |

*Source*: Osborne and Cormack, 1989.

Table 4.6 summarizes the data for O and A levels for 1975 and 1982. In general, examination passes obtained in Protestant schools are weighted towards the sciences, whereas examination passes obtained in Catholic schools are weighted towards humanities, languages, and arts subjects. These broad patterns are observable for both O and A levels. Looked at over time, however, these differences have shown some decline. In 1975, Catholic males gained 32% of their O level passes in mathematics and science compared with 42% of Protestant males, while 19% of Catholic females' and 27% of Protestant females' passes were gained in these subjects. By 1982 some expansion had taken place in the proportion of passes in mathematics and science, particularly in relation to Catholic females. However, most of this expansion was in biology rather than physics or chemistry. At A level the evidence of convergence seems a little more obvious with Catholic males, in particular, showing a movement into the sciences. Of potential significance for access to certain jobs in the labour market is the possession of an O level pass in what was previously called technical drawing. While Protestant and Catholic secondary schools recorded broadly similar proportions of boys passing in this subject (girls do not record any significant passes in this subject area) there were substantial differences between the representation of this subject in the profile of passes from grammar schools. While virtually all Protestant grammar schools recorded some passes in this subject, only a minority of Catholic grammar schools did so.

The second source of information comes from the School Census conducted by the Department of Education on an annual basis. Schools are responsible for completing standard forms covering basic information on pupil numbers and characteristics. In recent years this has included the A level subjects being studied. The A level subject combinations are collected under the headings of 'maths/science only', 'other subjects only', and 'combinations of the maths/science subjects with other subjects'.

The 1988 data from this source have been analysed by Osborne *et al.* (1989). Table 4.7 shows the A level subject combinations for Protestants and Catholics by gender. It can

TABLE 4.7. *Subject combinations of A level students, by religion and gender, 1986–7* (%)

|  | Science | Arts | Mix | Total |
|---|---|---|---|---|
| Protestant | 27.0 | 37.8 | 35.1 | 100 |
| Catholic | 18.7 | 48.5 | 32.8 | 100 |
| Protestant male | 36.8 | 26.7 | 36.4 | 100 |
| Protestant female | 18.4 | 47.7 | 33.9 | 100 |
| Catholic male | 26.6 | 40.1 | 33.2 | 100 |
| Catholic female | 12.0 | 55.6 | 32.4 | 100 |

*Note*: Totals may not equal 100 due to rounding.
*Source*: Osborne *et al.*, 1989.

be readily seen that Protestants are much more likely to have science combinations than Catholics and that, when gender is considered, this pattern is repeated both for males and females. Comparing grammar schools across Board areas shows Catholic boys' combinations to be more towards arts combinations for all areas. The strongest contrast between Protestant and Catholic boys occurs in the Western Board area where

TABLE 4.8. *Schools and science A levels: highest and lowest proportions, 1986–87*

| School | Lowest % of pupils studying all science subjects | School | Highest % of pupils studying all science subjects |
|---|---|---|---|
| Z | 4.4 | A | 49.0 |
| Y | 7.7 | B | 48.8 |
| X | 8.2 | C | 43.9 |
| W | 8.8 | D | 43.0 |
| V | 10.8 | E | 40.7 |
| U | 12.0 | F | 38.7 |
| T | 12.7 | G | 38.5 |
| S | 14.4 | H | 34.7 |
| R | 14.8 | I | 36.2 |
| Q | 15.3 | J | 35.1 |

*Source*: Osborne *et al.*, 1989.

23.2% of Catholic boys in grammar schools are studying all-science combinations whereas 39.6% of Protestant boys are doing so. Thus, while Catholic boys in grammar schools represent 64.1% of boys studying in grammar schools in the Western Board area they represent 51% of boys studying all-science A levels.

This analysis can be taken a stage further by examining patterns at the level of the individual school. In Table 4.8 are shown the ten grammar schools which have the lowest proportions taking all-science combinations and the ten grammar schools which have the highest proportion of those taking science A levels. All of the schools in the low science category are single-sex female schools and eight of the ten schools are Catholic. All but one of these Catholic schools are in the Western or Southern Board areas, the more rural parts of Northern Ireland. Amongst those schools with the highest proportions of those with science combinations, eight are Protestant schools (three of which are co-educational) and two are Catholic boys' schools. There is no geographical distinctiveness to the location of schools in this category.

This assessment of subject combinations at A level provides confirmatory evidence of the existence of a curriculum difference in the recent and contemporary outputs of the two school systems in Northern Ireland. The next piece of evidence to be examined relates to the extent and nature of qualifications in the adult population.

## QUALIFICATIONS AND THE ADULT POPULATION

The main source of information to establish the characteristics of the qualified population is the population census. In Table 4.9 are data from the 1971 and 1981 population censuses. They suggest that at the highest levels of qualification, as measured by the census, there are relatively small differences in the proportion of Protestants and Catholics with these qualifications. However, significant differences do emerge at this level of qualification in relation to subjects. Data from the 1981 census reveal that proportionately twice as many Protestants are qualified in science and technology while Catholics record a

TABLE 4.9. *Population aged 18+ with qualifications (at higher levels) 1971 and 1981, and subjects for 1981*

| | Catholic | | Protestant | | Catholic | Protestant |
|---|---|---|---|---|---|---|
| | M | F | M | F | | |
| % qualified 1971 | 4.6 | 7.3 | 7.6 | 6.9 | 6.0 | 7.2 |
| % qualified 1981 | 6.9 | 9.3 | 8.9 | 8.0 | 8.1 | 8.4 |
| Subjects 1981 | | | | | | |
| Education | 22.6 | 32.3 | 8.7 | 29.0 | 28.4 | 18.8 |
| Health | 15.3 | 44.9 | 12.2 | 41.3 | 33.0 | 26.7 |
| Science and technology | 22.8 | 3.4 | 38.4 | 5.5 | 11.2 | 22.0 |
| Social administration and business | 20.1 | 6.7 | 24.2 | 7.9 | 12.1 | 16.1 |
| Vocational | 2.1 | 1.7 | 3.6 | 3.3 | 1.9 | 3.5 |
| Arts and language | 17.2 | 10.8 | 12.9 | 13.0 | 13.4 | 13.0 |

*Note*: 'Science and technology' includes science, technology, engineering, agriculture, and veterinary science.

*Source*: Osborne and Cormack, 1987.

much higher representation in health and education. There are interesting contrasts when gender is also considered. Both male and female Catholics are strongly represented in education and health whereas for Protestants only females have a high representation in these subjects. The significance of education in the Catholic subject profile demonstrates the important role the existence of a separate Catholic education system has had for the employment of Catholics. The significance of these differences has undoubtedly grown with the 'tightening bond' in recent years.

## SUBJECT BALANCE: EXPLANATIONS

Several explanations have been advanced to try to explain the differences in curriculum balance between the two school systems. These explanations are complementary rather than mutually exclusive and remain possible rather than proven, with research to evaluate them fully still to be undertaken.

The first explanation points to different financial arrangements in Northern Ireland for the capital funding of schools. Currently all maintained schools (Catholic primary and secondary schools) and voluntary grammar schools (all Catholic and about two-thirds of Protestant grammar schools) receive 85% of capital costs. This figure was raised from 65% in the late 1960s and as a result, a considerable financial burden was placed on Catholic schools and the Catholic community to provide a network of schools able to offer a broad curriculum. During the 1950s and 1960s this was demonstrated, as Figure 4.2 shows, in a slower rate of school building in the Catholic post-primary sector (Osborne *et al*, 1989). It may well have been the case that the provision of expensive school laboratories was inhibited by these financial arrangements. In the light of the awarding of 100% capital funding and priority to the new category of 'integrated' schools and government proposals laying down a core curriculum for all schools (HMSO, 1989*a*) the level of capital funding of Catholic schools is returning to the political agenda (see below).

The second factor which has been identified relates to the ethos in Catholic schools. The rationale for a separate Catholic

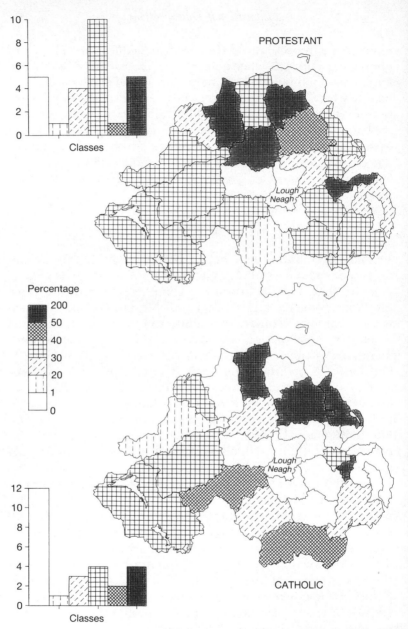

FIG. 4.2. Ratio of grammar school places to primary places by religious affiliation of school and District Council area in Northern Ireland

*Source:* Osborne *et al.*, 1989.

school system emphasizes the importance of moral development and pastoral concerns (Loughran, 1987). The direct
involvement of clergy and members of religious orders in
teaching, although now significantly reduced, may impart an
ethos which results in a less scientific orientation. There is some
evidence that head teachers in grammar schools in the two
systems conceive their roles differently, with Catholic prin-
cipals emphasizing their pastoral role while dealing with other
responsibilities through delegation, whereas Protestant heads
regard themselves as managers (Cormack, Miller, Osborne,
and Curry, 1989). However, there has been no direct research
into the question of 'ethos' and its possible role in influencing
curriculum in post-primary schools (but see Murray (1985) on
the ethos in primary schools).

Patterns of employment in the labour market may also exert
an effect on subject balance. Research evidence demonstrates
that the employment and occupational profiles of the two
communities are significantly different. Catholics continue to
be underrepresented in the broad area of manufacturing and
particularly in engineering and technology (see Chapter 2).
The role of both direct and indirect discrimination in laying
down and perpetuating these patterns has been widely ac-
knowledged. In these circumstances even though education
itself may have been highly valued it would not be surprising if
the Catholic system gave a lower emphasis to science and
technological subjects. For many in the Catholic system there
would have been few obvious employment outlets for pupils
specializing in science subjects and hence little incentive to
study these subjects.

The final factor identified relates to the possible difficulties in
recruiting teachers. The recruitment of good staff in the science
area is a widespread problem. In circumstances where, in
general, staff are recruited from the products of the segregated
system, it may well prove more difficult for Catholic schools to
recruit and retain staff. Given the profile of subjects studied, it
is likely that Catholic schools experienced problems in recruiting science teachers.

## HIGHER EDUCATION: PARTICIPATION AND GRADUATES' EARLY CAREERS

Recent large-scale survey research has enabled a comprehensive picture to be obtained of both levels of participation in higher education and the early career experiences of graduates. The surveys were undertaken by postal questionnaire and achieved high response rates (for the details of the surveys see Osborne *et al*, 1988). In Table 4.10 is shown the representation of Catholics and Protestants in higher education entrants for the survey years along with mean A level score of entrants. Because of the problems associated with the 1981 census (see Chapter 2) it is not possible to construct accurate age-specific cohorts nor, therefore, age participation rates (APRs). Nevertheless, estimated APRs for the two groups for earlier years suggested that the Catholic rate approximated the Protestant rate (Osborne *et al*, 1983). Clearly, Catholic participation levels have continued to increase. The mean A level score of entrants was similar in 1973; with the subsequent expansion of participation in higher education, Catholic scores have become lower than those of Protestants. Catholics are far more likely to come from manual backgrounds (48%) than Protestants (25.2%). Protestants are more likely to leave Northern Ireland to study while Catholics are more likely to remain there.

The surveys of undergraduate participation also collected data concerning the early labour market careers of Protestants and Catholics (Osborne and Cormack, 1989*b*; Miller *et al*, 1990). Overall, Protestant graduates are more likely to have secured a job commensurate with having a degree. Catholics were more likely to be unemployed or not working or to be working in jobs of a non-graduate type. More detailed breakdowns of economic activity demonstrated that there were significant differences in the type of work pursued by Protestant and Catholic graduates both for those employed in Northern Ireland and for those employed in Britain. Table 4.11 records the detailed occupational breakdown of the activities of 1979 entrants as graduates in 1985. The professions show the most marked differences between Protestants and Catholics. Taking

TABLE 4.10. *Higher education entrants by religion and mean A level points score, 1973, 1979, and 1985*

| | Protestant | | | Catholic | | |
|---|---|---|---|---|---|---|
| | N | % | Mean A level points score | N | % | Mean A level points score |
| 1973 | 863 | 68.9 | 9.9 | 405 | 31.1 | 9.4 |
| 1979 | 1,570 | 60.7 | 9.4 | 1,034 | 39.3 | 8.3 |
| 1985 | 1,966 | 56.2 | 9.9 | 1,534 | 43.8 | 9.0 |

*Source:* Osborne and Cormack, 1989.

TABLE 4.11. *Present economic activity of graduates by location and religion* (%)

| Activity | Present location | | | |
|---|---|---|---|---|
| | Northern Ireland | | Britain | |
| | Protestants | Catholics | Protestants | Catholics |
| Managers/employers, large enterprises | 2.4 | 0.7 | 2.8 | 0.0 |
| Managers/employers, small enterprises | 6.5 | 6.5 | 6.8 | 4.8 |
| Legal professionals | 2.5 | 5.0 | 3.1 | 3.2 |
| Medical professionals | 8.8 | 6.0 | 6.5 | 3.2 |
| Business support professionals | 4.9 | 2.9 | 4.5 | 1.6 |
| Engineering professionals | 5.0 | 2.2 | 10.2 | 7.1 |
| Other professionals | 4.0 | 4.1 | 5.4 | 5.6 |
| Teachers | 8.1 | 7.5 | 5.4 | 3.2 |
| Social service semi-professionals | 3.7 | 4.1 | 3.7 | 5.6 |
| Nursing and health support | 2.9 | 1.7 | 3.4 | 7.1 |
| Computing and technicians | 3.8 | 4.3 | 6.2 | 6.3 |
| Other intermediate non-manuals | 5.6 | 6.3 | 7.6 | 7.9 |
| Routine non-manuals | 12.1 | 10.6 | 7.4 | 11.1 |
| Manual and farm | 1.9 | 3.1 | 2.5 | 2.4 |
| Postgraduate education | 20.3 | 24.3 | 19.3 | 18.3 |
| Not working/unemployed | 7.5 | 10.8 | 5.1 | 12.7 |
| TOTAL | 100 | 100 | 100 | 100 |
| N | 679 | 416 | 353 | 126 |

*Source:* Miller *et al.*, 1990.

those living in Northern Ireland first, Protestants are more likely to be managers in large establishments, and professionals in medicine, engineering, and business. On the other hand Catholics are markedly more likely to be in professional occupations in law. Thereafter, variations in occupations are quite small. For those domiciled in Britain, Protestants are also more likely to be managers/employers, and professionals in engineering, medicine, and business. Catholics, on the other hand, were more likely to be in the semi-professions in the health and social services and in routine non-manual jobs or unemployed.

Overall, when the early careers of graduates are analysed, differences between Protestants and Catholics were significantly related to the courses undertaken in higher education. As we have seen, subject differences are evident at the point of entry to higher education. Religion, *per se*, did not therefore exert a significant effect on economic activity when such factors are taken into account. The evidence suggests no religion-specific pattern of disadvantage: a Protestant lower attainer who undertook a non-science course was similarly disadvantaged.

## QUALIFICATIONS AND UNEMPLOYMENT

A key element of the meritocratic assumptions about the value of educational qualifications in the labour market is that the possession of any particular qualification or set of qualifications should broadly result in similar patterns of occupational success for different social groups. If one particular group is significantly less able to utilize educational attainment in this way than other groups then it implies that non-meritocratic criteria may be determining job access and success. A key measure of this is the gaining of a job. From the Continuous Household Survey (CHS) it is possible to examine the unemployment rates for those with qualifications and those without qualifications. The data in Table 4.12 show the rates of unemployment as recorded in the 1983–4 CHS.

For both Protestants and Catholics the possession of a qualification lowers the rate of unemployment, but Catholics

TABLE 4.12. *Unemployment rates for those with a qualification and those without (population 16+ and economically active), Continuous Household Survey 1983–4*

|                    | No qualifications | Qualifications |
|--------------------|-------------------|----------------|
| Protestant males   | 20.3              | 10.8           |
| Protestant females | 12.1              | 9.4            |
| Catholic males     | 42.2              | 26.1           |
| Catholic females   | 18.0              | 15.5           |

*Source*: Osborne and Cormack, 1987.

with a qualification are more likely to be unemployed than Protestants. This general picture is congruent with the results from the Youth Training Programme (YTP) cohort study where it is revealed that those Catholics with qualifications have been less likely to convert these credentials into a full-time job than their Protestant contemporaries (see Chapter 5).

However, as we have discussed, most of the employment disadvantage experienced by recent Catholic *graduates* is a result of differences in discipline studied; there are no particular disadvantages experienced by Catholics when compared to Protestants who have studied the same subjects. It may well be, therefore, that more modestly qualified Catholics, that is those with a few CSEs or one or two O levels (now GCSEs), are particularly disadvantaged in the labour market but that Catholic graduates experience no disadvantages beyond those associated with the characteristics of their educational credentials. A partial explanation could stem from the personnel practices relating to different types of job and employers. Those jobs sought by graduates are more likely to be with employers operating formal personnel procedures, while those sought by the less qualified academically tend to rely more on informal methods and to be with employers who may be more prone to condoning both direct and indirect discrimination. It is also possible that graduates are more geographically mobile and willing and able to take jobs across a wider geographical area.

## CONCLUSIONS

What are we to make of the differences in educational qualifications between Protestants and Catholics for fair employment policy in Northern Ireland? A number of points can be made.

Recent school-leavers, graduates, and the existing qualifications of the adult population continue to show a bias in the Catholic profile away from science and technological subjects. Catholics are clearly going to be at a disadvantage where jobs requiring such qualifications are concerned. It is not that Catholics do not have these qualifications but that, proportionate to their representation in the qualified population, Catholics are significantly under-represented in the science and technology areas. There is little doubt that for the forseeable future, jobs demanding technological skills will be in high demand. On our evidence, it would seem that Catholics and especially Catholic females are poorly placed to take advantage of these opportunities and Protestants, particularly Protestant males, are well placed.

From the evidence we have produced, there is a need for the Catholic school authorities to give serious consideration to the curriculum balance in their schools. The intention of the Department of Education for Northern Ireland to introduce a 'core' set of subjects for all schools which will include science and technological subjects may well have the *long-term* effect of increasing the representation of science in Catholic schools. It is also clear, however, that such a change will have major implications for schools since it will require significant increases of expenditure on new laboratories and purpose-designed technology suites and staffing. Since all Catholic schools under current arrangements must raise 15% of capital costs, a major financial burden will fall on the Catholic authorities and population. By the same token, proposals to introduce individual school budgets on a new formula basis rather than the existing historical budgets may mean significant enhancement of some schools' revenue budgets as they move from an arts-dominated, and therefore cheaper curriculum, to one more balanced between the arts and sciences. Since the state is seeking such changes, should the state pay? This question goes to the heart of the issue of control. If 100% capital funding of Catholic schools was offered, would Catholic schools be asked to concede an element of control in return, and if that was the nature of the offer would the Catholic authorities accept? (See the House of Commons debate of the Education Reform (Northern Ireland) Order, where a government minister seemed to suggest that just such an offer could be available (*Hansard*, 1989).) The issue is widened further if the position of the Catholic laity is considered. Would the Catholic community adopt the same position on these matters as the church authorities?[1]

There may well be a further knock-on consequence of this under-representation in science. Increasing numbers of employers are making use of selection tests in order to make more effective and objective decisions. Tests which seek to explicitly measure mechanical/scientific knowledge and/or aptitude, on the basis of our evidence, may well produce lower Catholic mean scores. Indeed, the third report of the Equal Opportunities Unit of the Northern Ireland Civil Service (1988) reveals major differences in the performance of Protestants and Catholics, to the latter's disadvantage, in the general UK tests for civil service entry to administrative and executive officer grades. The report further indicates that it has not been able to identify the factors producing these results but notes that those with a scientific background tend to do better on all aspects of the tests (Chapter 7; Osborne, 1990). What is not entirely clear, however, is whether such a pattern of results would reflect real potential relative to acquired knowledge. How should employers respond to such results? It would seem that they will need to respond in much the same way as in the case of similar

---

[1] The increase in the funding of Catholic schools in Ontario makes an interesting comparison. The awarding of full funding of Catholic schools announced by the Ontario provincial government in 1985 has also involved the ending of the right of the Catholic authorities to appoint only Catholics as teachers after 1995 (see Walker, 1986; Osborne, 1989). However, a more direct example, within the UK, is Scotland where Catholic schools, by being absorbed into the public system, have received 100% revenue and capital funding since 1918 and where the Catholic church has retained, for example, a level of control over the appointment of teachers which it regards as satisfactory (Treble, 1979; Cormack, Gallagher, and Osborne, 1990).

results for women. In a study conducted for the Equal Opportunities Commission in Great Britain it is suggested that users of tests should select tests carefully, avoiding strict cut-off points except where there is a clear relationship between test score and job performance, prepare candidates for testing, look for differences between groups, and support tests with other information (Pearn *et al.*, 1987). In the Northern Ireland context very little work has been reported in this field. There would seem to be a major opportunity and responsibility on the local branch of the British Psychological Society to provide professional guidance and assistance to employers using psychological tests in Northern Ireland.

In the context of the new fair employment legislation, with its emphasis on affirmative action which may specify goals and associated timetables, it is clear that guidance to employers will need to reflect these educational differences in order that realistic goals can be set.

The final point to make in this regard is that although there are differences in the qualifications of Protestants and Catholics, a significant proportion, if not a majority, of jobs in the labour market do not require the possession of academic or academic-related qualifications for access. It will be a major feature of the implementation of the new fair employment legislation that all direct and indirect barriers to equal competition for all jobs must be removed by employers.

The role of educational qualifications in accounting for employment differences between Protestants and Catholics is complex. There is no doubt that real differences exist and that the achievement of greater equality of opportunity will be enhanced by the strengthening of science and technological subjects in Catholic schools that may eventually result from the introduction of the 'core curriculum'. However, it must be a major goal of the new fair employment legislation to ensure that those who secure academic credentials are equally able to translate those achievements into labour market success.

# 5

# Religious Affiliation and the Youth Training Programme

*Liz McWhirter, James Gillan, Ursula Duffy,*
*and Denise Thompson*

## INTRODUCTION

The issue of fair employment between the two main community groups in Northern Ireland, Protestant and Catholic, has been the focus of attention by politicians, policy makers, and researchers in recent years (Cormack, Osborne, and Thompson, 1980; Darby and Murray, 1980; Osborne and Cormack, 1987, 1989*b*). While the question of equality of opportunity in training for employment has received some consideration with regard to males and females in Northern Ireland (see, for example, McWhirter *et al*, forthcoming), little direct attention has been given by researchers to equality of opportunity in employment training schemes with respect to the religious affiliation of the trainees. An equal opportunities code relating to sex and disability is already well advanced for YTP participants and the advent of the Fair Employment (Northern Ireland) Act of 1989 highlights the importance of addressing the issue of religious affiliation in youth training.

This chapter will consider the topic of equal opportunity for Protestants and Catholics on the Youth Training Programme (YTP), the Northern Ireland equivalent of the Youth Training Scheme in Britain. The chapter is based upon evidence derived from a major longitudinal investigation of young people in the Province, the YTP Cohort Study, field-work completed in 1987.

Government-sponsored special measures to combat unemployment have spanned a variety of types of schemes but youth training has become the policy response for the under-18-year-olds. The key transition policy for young people in Northern

Ireland since 1982 had been the Youth Opportunities Programme (YOP). The Programme was designed to contribute to the development of the Northern Ireland economy by laying the foundations of a skilled, flexible work-force and, as originally conceived, aimed to provide an integrated training and education programme for all 16- to 18-year-olds. As such, the two-year Programme set out to offer a full-time combination of vocational training to those 16- and 17-year-olds not yet in employment, employment with training opportunities to 17-year-olds, and increased vocational preparation to those young people who remain in full-time education. In practice, however, YTP has operated essentially through the provision of a variety of training and work schemes for unemployed 16- and 17-year-old school-leavers, who are the group most in need. Full-time training places on YTP are offered to all 16-year-old school-leavers through partnerships organized largely by Training Centres, Community Workshops, Further Education Colleges, and employers. The major element in the second year provision is YTP Workscheme, which is a 'real' job with training.

YTP aims to strengthen the economy of Northern Ireland and to enhance the personal development of its participants. Within the context of the equal opportunities debate in Northern Ireland, it is perhaps of particular interest to establish how well YTP is faring in what might be regarded as a pivotal role in the employment sector. YTP is well positioned to provide a unique contribution in helping to shape the characteristics of the Province's emergent work-force. Indeed, with the recent changes in the social security system for under-18-year-olds, the impact of YTP in introducing young Catholics and Protestants to the labour market is arguably of even greater importance. Since the autumn of 1988, YTP has in effect become obligatory for the vast majority of those young people who do not obtain a job immediately upon leaving school at age 16 or 17. Promotion of equal opportunities practices within the control of the Programme's co-ordinators should obviously, therefore, be of prime concern. It is also important to recognize, however, that the Programme was not specifically established to compensate for any imbalances which might exist in the wider labour market. The YTP Cohort Study provides a unique vantage-point from which to examine some of the equal opportunities issues associated with the Programme.

As part of the continuing evaluation of the Programme, the Departments responsible for YTP commissioned the Policy Planning and Research Unit (PPRU) of the Department of Finance and Personnel in 1984 to undertake an interview-based investigation of the routes taken by young people on reaching school-leaving age (16) and in the three to four years thereafter. The target population for the study was young people who became eligible to leave school in the academic year 1983–4. A 10% sample of this group was selected for inclusion in the study (N = 2,890). While sampling was proportionally stratified according to the type of school attended (secondary or grammar), the school location (area board), and management type (controlled or maintained), the sample was weighted in favour of secondary school pupils in order to reflect the wider range of vocational choices likely to be made by pupils attending such schools (see McWhirter *et al.* (1987) for further details of the sampling). Any attempt to generalize directly from the study must, therefore, be circumscribed by this in-built bias and the sampling frame described.

Field-work for the study, which occurred in five sweeps of data collection, spanned approximately three and a half years. It commenced in April 1984 when the young people were in their final term of compulsory schooling and ended in October–December 1987 when the young people averaged 19 years of age (see McWhirter *et al.*, forthcoming, for details of field-work). Approximately two-thirds of the original sample were still fully co-operative at the final stage of the study.

The stage-five sample comprised 56% Protestants and 42% Catholics (the remainder were not classified). There were the same proportions of males and females in the two religious groups, but the Protestant group came from more advantaged home backgrounds and tended to be somewhat better qualified than the Catholic group (see McWhirter, 1989, for details).[1]

---

[1] Young people participating in the study were not asked about their religious affiliation directly, rather it was assigned on the basis of the school

## THE PRESENT CONCERN

The chapter will examine the issues of religious group differences in relation to the YTP by considering data from the study which concern three different dimensions of the equal opportunity debate: equal access, equity of treatment, and equality of outcome. The provision of information about the Programme to Protestant and Catholic pupils prior to their leaving school, the ratios of Protestant and Catholic participants on the Programme and on different YTP schemes, and the reasons members of the two groups have for joining YTP are all relevant to the question of equal access.

The question of similarity of achievement is a much more difficult and far-reaching issue than equality of access. As Stevenson *et al.* have argued, the operation of practices which afford equality of opportunity may not in fact lead to equality of outcome. 'Imbalances [in employment] may reflect intractable non-discriminatory reality, e.g. the effects of persistent group differences in job-related educational qualifications' (Stevenson *et al.*, 1988: 259–60).

A comparison of the proportions of Protestants and Catholics who achieved qualifications while on the Programme is one appropriate criterion to assess equality of outcomes for the two groups. The value of achieving a qualification during YTP training in order to enhance the chances of the young person securing employment afterwards is supported by the finding (Faloona *et al*, 1988) that qualifications were viewed as an important selection criterion by 44% of employers in Northern Ireland. Differences in the opinions and levels of satisfaction held by the two groups about their YTP experiences are also pertinent to the issues of similarity of achievement, as well as providing insights into the question of equity of treatment on the Programme.

attended in the fifth form (see Ch. 4). For the purpose of classifying the young people on the basis of attainment, a composite educational performance score, attributing different weights to different levels of academic and vocational qualifications (DENI) was used. Social background was also determined by using a composite measure, derived from Osborn *et al.* (1984), which included parents' educational status, social class of father's job, and the nature of house tenure.

A more objective and more commonly used measure of the efficacy of the Programme which is applicable to the topic of equality of outcomes is post-YTP employment rates. While YTP is a training measure and not an employment scheme, it is reasonable to expect trained young people to be more employable than their non-trained counterparts who did not go directly from school to work after leaving school, particularly in an area of high unemployment such as Northern Ireland. This training 'advantage' might also be expected to apply equally to Protestant and Catholic YTP trainees relative to their co-religionists who have not taken part in the Programme. On the other hand, post-YTP employment rates are outside the direct control of the Programme. Catholic and Protestant young people may only be returned to a situation in the wider labour market where the varying socio-economic and other factors underlying their ability to find work continue to operate. With these constraints in mind, consideration of the similarity of outcomes with respect to employment patterns will be examined through the use of multivariate analyses. In addition, these analyses will take into account important differences in a number of key personal characteristics of the Protestants and Catholics in the study. The simultaneous examination of relevant variables such as educational performance and home background alongside religious affiliation and YTP experience leads to a more thorough analysis of the complex set of interactions and processes which together contribute towards equal or unequal outcomes in the labour market.

For ease of presentation, this chapter will consider the evidence in chronological order as far as a young person's YTP career would be concerned. Thus, we shall begin by looking at awareness of and attitudes towards YTP among pupils prior to school-leaving age and then move on to an exploration of participation rates on YTP and why young people joined the Programme. Finally, the chapter will examine some of the outcomes of YTP participation: opinions of and satisfaction with YTP, qualifications obtained whilst on the Programme, post-YTP employment rates, and types of jobs obtained.[2]

[2] Many of the results are presented in summary form only. Further details are available from the authors.

## AWARENESS OF YTP AMONG PROTESTANT AND CATHOLIC PUPILS

The extent to which young people are aware of YTP in the first instance, one could presume, is one of the factors which will influence their decision to enter the Programme. When interviewed in April 1984 (their final term of compulsory schooling), just over two-thirds of the young people indicated that they knew at least something about YTP. More secondary school pupils than grammar school pupils were informed about the Programme but there were no marked differences between the religious groups, either overall or within the secondary and grammar school groups.

Given that the religious groups appeared to be equally well informed about YTP while approaching the end of their compulsory schooling, any differences in the uptake of YTP among them would appear to be a consequence of factors other than overall level of knowledge about the Programme. Young people's sources of information on the Programme are also of relevance in assessing the extent to which it was being equally promoted among Protestants and Catholics. Whilst careers teachers (64%), friends (54%), and careers officers (47%) were the three most frequent sources of information on YTP for both groups overall, there were notable differences between the two religious groups and between the school types, secondary and grammar.

Within the grammar school sector more of the Catholic pupils had learned about the Programme from careers teachers, careers officers, friends, and family, while the media—television and newspapers—had been more instructive for the Protestant pupils. Within the secondary school sector, the main target group for YTP schemes, Catholics were more likely to have learned informally about YTP from their friends and their families whilst more of Protestants had been informed officially about YTP from careers officers. These different sources of information on YTP for the secondary school pupils are notable but are balanced somewhat by the finding that the principal source of information on YTP, the careers teacher, was equally active in both the controlled

(Protestant) and maintained (Catholic) secondary schools (72% and 71% respectively).

The above findings suggest that it is unlikely that any imbalance in the participation rates of YTP among the religious groups was due to the sources of knowledge about the Programme—unless, of course, the two groups were being given different messages about YTP. It is, therefore, of interest to explore the opinions the young people held towards the Programme.

## ASPIRATIONS AND ATTITUDES TO YTP

An appraisal of the young people's attitudes to YTP must be set in the context of their views on other economic activities. Analyses of post-16 preferences expressed during the fifth form reveal no statistically significant differences in the post-compulsory school choices stated between the religious groups. Forty-five per cent of the Protestants and 48% of the Catholics at age 15 wished to remain in full-time education, while 53% of the Protestants and 49% of the Catholics preferred to leave school at 16 and get a job. A very small proportion (2%) said they would choose YTP in preference to continuing in education or finding employment. This small proportion reflects the general perception that the Programme is targeted mainly, in practice, at the unemployed. In addition, there were no significant differences in commitment to work (Warr *et al*, 1979) between the two groups; attitudes were generally favourable. There were also no religious differences in the importance attached by the cohort at age 17 to the role of different factors in the job search (McWhirter *et al*, 1987). The data thus suggest that attitudes to work for both the Catholics and the Protestants were quite similar.

In sharp contrast, however, there were significant differences in attitudes to YTP, as measured by a 12-item scale (reliability coefficient—Cronbach's Alpha = 0.8) administered in the fifth form. The Protestant pupils were more positive towards the Programme than the Catholics. In particular, more Catholics than Protestants thought that the training allowance was not enough and that YTP was only a way of taking young people 'off the dole' for a while. Furthermore, the reasons given by trainees for joining YTP after reaching age 16 also suggest that the Protestants were somewhat more optimistic about the value of the Programme than were the Catholic trainees. Although the most frequent reason given for joining YTP for both groups was that it was preferable to being unemployed, more Protestants than Catholics gave this as their most important reason for joining the Programme. These differences in attitudes perhaps reflect the different unemployment rates in the two communities (see below). Also, while both groups gave as their main reason for joining the Programme 'wanting to be trained', more Protestants than Catholics stated that they joined YTP because of the desire to be trained (McWhirter *et al.*, 1987).

The data thus indicate that any discrepancies in Protestant and Catholic participation rates in YTP could perhaps be due, at least in part, to the observed differences in views towards the Programme. A more influential cause, however, was likely to be whether or not Protestants and Catholics secured a full-time job on leaving school. YTP is, in effect, for the young *unemployed* labour market entrant and in practice the client group for YTP schemes comprises those who do not move directly from school to employment.

## PROTESTANT AND CATHOLIC PARTICIPATION RATES IN YTP

All young people leaving school at age 16 are now guaranteed two years of YTP training. Since the Programme appears to have been equally well promoted among Protestant and Catholic school pupils, it is of interest to ascertain whether the two groups have been equally likely to join the Programme. In order to assess Protestant and Catholic participation rates in YTP it is instructive to examine the economic activity patterns of the total cohort at age 16 before examining those of the labour market entrants.

Two to six months after reaching school-leaving age more than half (52%) of the young people in the YTP Cohort Study overall, but considerably fewer Protestants than Catholics, were still pursuing full-time education (48% and 57% respect-ively). YTP was the next most likely post-16 destination for the young people overall (22%), with almost half (45%) of the school-leavers having joined the Programme. Catholic labour market entrants were more likely than their Protestant counterparts to have entered YTP at this time (51% and 41% respectively).

Although the joining rate for YTP among the Protestants and Catholics in the labour market at age 16 differed markedly, data from later stages of the study indicate that the same proportions of the two groups participated in the Programme at some stage throughout the three and a half year period of the study (33% Protestants and 34% Catholics).

Why the differences in YTP participation rates among the Protestant and Catholic labour market entrants at age 16 occurred is of interest, especially as we have noted that the Catholics held less favourable opinions of YTP when they were still at school. One obvious reason relates to the chances of the two groups securing employment on leaving education. The data indicate that Protestant labour market entrants had fared better in the job hunt than Catholics (43% of the Protestants secured full-time work shortly after leaving school compared with 33% of Catholics). It is thus not surprising that a smaller proportion of the Protestant school leavers were in YTP. Therefore the net result of more Catholics staying in education beyond age 16 and more Protestants obtaining jobs imme-diately on leaving school was that for Protestant and Catholic groups *overall* there was equal uptake of YTP over the period of the study.

The choice of YTP schemes also differed substantially be-tween the religious groups (Table 5.1). At age 16, the most usual scheme for the Protestants interviewed was the college-based scheme, while for the Catholics Training Centres were most frequent. One year later there were still marked dif-ferences in the types of scheme joined by the two groups. Protestants, both male and female, were more likely than their Catholic counterparts to be in Workscheme (which is a real job where the young person is an employee rather than a trainee), while Catholics, both male and female, were about

TABLE 5.1. *Proportion of young people on each type of YTP Scheme at ages 16 and 17* (%)

| Scheme entered | Age 16 | | | Age 17 | | |
| --- | --- | --- | --- | --- | --- | --- |
| | All | Protestant | Catholic | All | Protestant | Catholic |
| Workscheme* | — | — | — | 32 | 36 | 27 |
| College-based | 29 | 38 | 22 | 7 | 6 | 7 |
| Training Centre | 28 | 28 | 29 | 14 | 16 | 13 |
| Work Preparation Unit/Community Workshops | 19 | 15 | 22 | 20 | 13 | 26 |
| Employer-based scheme | 9 | 8 | 9 | 8 | 8 | 7 |
| Youthways | 7 | 6 | 9 | 3 | 2 | 4 |
| Young Help | 2 | 2 | 1 | 3 | 2 | 4 |
| Enterprise Ulster | 2 | 2 | 3 | 3 | 3 | 4 |
| Youth Community Projects† | — | — | — | 6 | 8 | 4 |
| Other | 3 | 1 | 6 | 6 | 7 | 5 |
| TOTAL | 100 | 100 | 100 | 100 | 100 | 100 |
| N | 688 | 371 | 302 | 263 | 122 | 140 |

*Note:* Totals may not add to 100 due to rounding.

* Available only to 17-year-olds.

† Includes National Trust, Attachment Training Scheme, and cases where the type of scheme was not identified. Available only to 17-year-olds.

twice as likely as Protestants to be on Community Workshop schemes.

The explanation for the variations in take-up of both first- and second-year schemes between the two religious groups is likely to be strongly related to the residential segregation of the two religious groups in the sample, and to differences in the location of employers and the siting of particular types of YTP schemes throughout the Province. The sampling design of the YTP Cohort Study renders it difficult to analyse geographical factors. Because of the relatively small numbers involved in the YTP schemes, for example, it is difficult to ascertain the extent of religious group segregation between or within particular centres. Whilst a new PPRU longitudinal study of YTP participants which is currently being planned may include equal opportunities in its brief, the Whyte *et al.* (1985) study of first-year trainees on the initial year of the Programme revealed that religious segregation did exist: some projects were almost entirely Protestant while others were composed almost entirely of Catholics.

## SATISFACTION WITH YTP

Having made the decision to join YTP, were Catholics and Protestants equally content with the Programme? Given the imbalances between different types of scheme it is to be expected that the opinions of the two groups would vary with respect both to different specific facets of their YTP experiences and to global satisfaction. The views of the participants on their YTP experience at ages 17, 18, and 19, are of relevance to the issues of equity of treatment and equality of outcome for the two groups.

Despite the differences noted earlier in attitudes towards YTP among the young people while still at school there were few differences between the views held by the religious groups about their *actual* experience of training schemes at age 17 (Table 5.2). The groups were equally—and well—satisfied with all aspects except for the further education component and the training allowance. Catholic trainees, however, were somewhat less content with both of these aspects and were also less happy than the Protestants about the assistance that they had received with numeracy. Two years later, the views of the two groups remained equally positive about many aspects of their YTP training but Protestants held more favourable opinions than Catholics about a range of issues. The one issue which appeared to be valued by more Catholics than Protestants concerned relationships with people, including those in authority. In contrast, the Catholic Workscheme employees were much more positive at both ages 18 and 19 than their Protestant counterparts about the help received from Workscheme. Similar proportions of Protestants and Catholics (46% and 43% respectively) indicated that they were given the opportunity to acquire a vocational qualification while on the Programme. Of this group, slightly more Catholics than Protestants availed themselves of the opportunity to acquire a vocational qualification (87% and 81% respectively).

### EMPLOYMENT PROFILES

In examining the extent to which Catholics, as such, may be disadvantaged in the youth labour market it is important both to compare the proportions of the two groups in employment and to carry out a broader examination of the types of jobs found by young Protestants and Catholics.

Employment profiles for Protestants and Catholics are illustrated in Figures 5.1 and 5.2 in relation to the other main education/labour market activities (ages 16–19) in the study. This information, derived from a monthly 'diary' completed by the cohort, indicates that a higher proportion of Catholics than Protestants remained in full-time education and slightly more of the Catholics were in YTP in the earlier stages of the study. In general, Protestants gained jobs at a faster rate than Catholics and the difference in unemployment rates of the two groups also increased with age.

Examination of the different proportions of the religious groups at age 19 in each of the main destinations after compulsory schooling (Figure 5.3) confirms these trends. Overall, Protestants at age 19 were 1.8 times more likely than the

TABLE 5.2. *How YTP training and workscheme helped young people: reactions at ages 17, 18, and 19 (%)*

| | YTP Training | | | | | | Workscheme | | | | | |
|---|---|---|---|---|---|---|---|---|---|---|---|---|
| | Age 17 | | | Age 19 | | | Age 18 | | | Age 19 | | |
| | All | Prot-estant | Cath-olic | All | Prot-estant | Cath-olic | All | Prot-estant | Cath-olic | All | Prot-estant | Cath-olic |
| Make new friends | 93 | 92 | 93 | 90 | 89 | 90 | 87 | 84 | 90 | 84 | 84 | 84 |
| Learn new work skills/learn new skills | 90 | 89 | 89 | 81 | 81 | 79 | 80 | 79 | 81 | 77 | 72 | 86 |
| Learn to work as part of a team | — | — | — | 79 | 79 | 80 | 81 | 79 | 84 | 74 | 70 | 79 |
| Learn to accept criticism and advice | — | — | — | 78 | 79 | 80 | 81 | 79 | 84 | 74 | 70 | 79 |
| Learn to be punctual | — | — | — | 77 | 77 | 77 | 80 | 76 | 84 | 77 | 74 | 81 |
| Get on better with people/ people in authority | 84 | 83 | 84 | 77 | 73 | 81 | 80 | 75 | 85 | 75 | 72 | 81 |
| Improve self-confidence | 81 | 80 | 80 | 75 | 73 | 75 | 81 | 77 | 86 | 73 | 71 | 78 |
| Become more mature and responsible | — | — | — | 72 | 72 | 73 | 83 | 78 | 88 | 76 | 73 | 82 |

| | | | | | | | | | | | | |
|---|---|---|---|---|---|---|---|---|---|---|---|---|
| Work without close supervision | — | — | — | 71 | 71 | 70 | 86 | 83 | 88 | 79 | 75 | 86 |
| Become aware of strengths and weaknesses | — | — | — | 70 | 69 | 70 | 71 | 71 | 71 | 69 | 63 | 78 |
| Learn about health and safety at work | — | — | — | 69 | 72 | 66 | 56 | 54 | 60 | 51 | 52 | 48 |
| Find out what sort of job you wanted to do | 66 | 62 | 67 | 58 | 59 | 57 | 61 | 60 | 64 | 54 | 49 | 61 |
| Know how to look for a job | 58 | 57 | 55 | 52 | 56 | 48 | 43 | 38 | 49 | 34 | 32 | 36 |
| Get a job | 46 | 46 | 41 | 49 | 51 | 46 | — | — | — | 57 | 56 | 62 |
| Management of money | — | — | — | 44 | 43 | 46 | 60 | 58 | 60 | 53 | 52 | 56 |
| Obtain a qualification | — | — | — | 36 | 40 | 30 | 32 | 31 | 34 | 28 | 32 | 23 |
| Improve arithmetic | 31 | 34 | 24 | 30 | 32 | 28 | 35 | 34 | 37 | 32 | 36 | 27 |
| Develop computing skills | — | — | — | 28 | 31 | 26 | 17 | 19 | 14 | 21 | 19 | 23 |
| Improve reading and writing skills | 27 | 28 | 23 | 27 | 30 | 24 | 22 | 20 | 26 | 26 | 24 | 28 |
| N | 698 | 368 | 316 | 480 | 270 | 197 | 345 | 200 | 135 | 232 | 131 | 95 |

*Note:* — means not asked.

% of total cohort

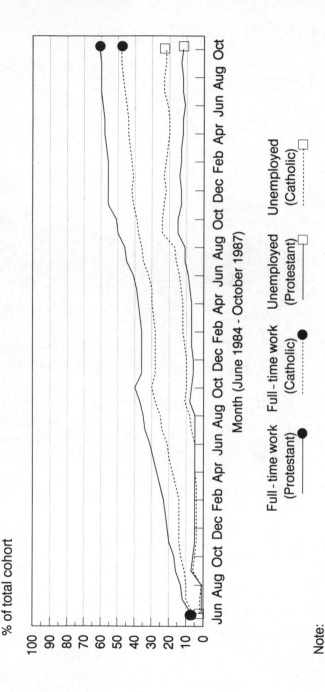

Jun Aug Oct Dec Feb Apr Jun Aug Oct Dec Feb Apr Jun Aug Oct
Month (June 1984 - October 1987)

Full-time work        Full-time work        Unemployed          Unemployed
(Protestant)          (Catholic)            (Protestant)        (Catholic)

Note:
Protestant (N=1075)
Catholic (N=774)

FIG. 5.1. Diary activities for Protestants and Catholics: work and unemployment. (Weighted N for total cohort)

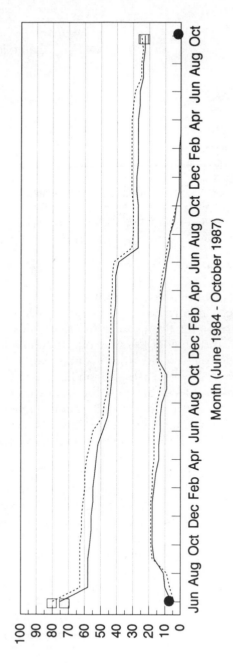

% of total cohort

100
90
80
70
60
50
40
30
20
10
0

Jun Aug Oct Dec Feb Apr Jun Aug Oct Dec Feb Apr Jun Aug Oct

Month (June 1984 - October 1987)

YTP/Workscheme    YTP/Workscheme    Full - time education    Full - time education
(Protestant)      (Catholic)        (Protestant)             (Catholic)

Note:
Protestant (N=1075)
Catholic (N=774)

FIG. 5.2. Diary activities for Protestants and Catholics: YTP/Workscheme and education. (Weighted N for total cohort)

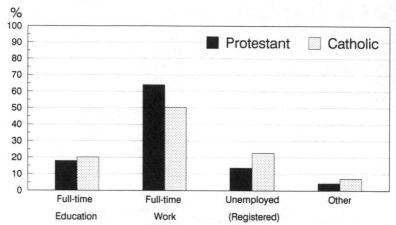

F IG. 5.3. Main activity of young people at age 19 by religious affiliation

Catholics surveyed to be in full-time work relative to the other activities described.

Of particular interest in the present chapter is the extent to which YTP operates as a mediating influence in determining different employment rates for Protestants and Catholics and in influencing the type and level of job they obtained. To this end, subsequent analyses will exclude those young people who went directly from school to work (with the job lasting at least six months) and will instead focus on school-leavers who were unable to find work within three months of leaving school or whose job on leaving school lasted less than six months. Within the operational context of YTP this is the population of young people to whom the Programme is directed and the most appropriate source for identifying a 'no YTP experience' comparison group. Thus the analyses focus on the Programme as being directed primarily at young unemployed school-leavers. The present concentration on the group of young people who did not proceed directly to employment contrasts with the previous reports of the study (e.g. McWhirter *et al.*, 1987; McWhirter, 1989), which had adopted the comprehensive school-to-work model that is suggested by the official description of the Programme, but not by the operation of YTP in practice.

The post-YTP employment rates of Protestants and Catholics in the study have been examined both descriptively and inferentially in order to explore as rigorously as possible the role of religious affiliation in obtaining full-time employment. In particular, the employment rates of those with YTP experience have also been compared with the employment rates of those who did not join the Programme.

When we compare the economic activities of the young people in the study at age 19 according to their activities at age 16 (Table 5.3) we see that twice as many of those who had joined YTP at age 16 were known to have secured full-time work three and a half years later as were known to be unemployed. In contrast, those who were unemployed after leaving school at age 16 were almost as likely to be out of a job as in work at age 19. It is noteworthy that the ratios of being in full-time work as opposed to being unemployed at age 19 for all of the Protestant groups is greater than for their Catholic counterparts. For example, in this subgroup, the Protestant who was in YTP at age 16 is nearly two and a half times more likely to be in work at age 19 than to be unemployed while the comparable ratio for the Catholics is one and a half to one. This difference is also starkly represented in Figure 5.4, which focuses only on the work and unemployment rates of those who joined the Programme at age 16.

YTP participants at age 19 tended to have a higher employment rate (70%—as opposed to being unemployed) than their counterparts who were unable to find a job within three months of leaving school and who did not join YTP (66% employment rate). While the relatively small numbers in each of the different schemes preclude detailed comparisons in employment rates between Protestants and Catholics at this level, there is some evidence (see Table 5.4) to indicate that Protestant Workscheme participants fared slightly better in obtaining a job than their Catholic counterparts. In examining the employment prospects of Protestant and Catholic YTP participants it is important to remember, however, that the groups differed in ways other than religious affiliation. As in the overall sample, more of the Catholic participants (66%) in YTP tended to be less well qualified than Protestant YTP participants (56%) and more Catholics in YTP (72%) also came from less advantaged

TABLE 5.3. *Main activity at age 19 according to main activity at age 16* (%)

| Activity at age 16 | Activity at age 19 | | | | | | | |
|---|---|---|---|---|---|---|---|---|
| | Full-time education | Full-time work | YTP/Work-scheme | Government Train-ing Scheme | Un-employed (Registered) | Part-time work | Other | Unknown |
| All | | | | | | | | |
| Full-time education (N = 1,435) | 23 | 35 | * | 1 | 8 | 2 | 1 | 30 |
| Full-time work (N = 490) | * | 59 | 0 | * | 7 | 1 | 2 | 30 |
| YTP (N = 589) | 1 | 42 | 0 | 1 | 20 | 1 | 2 | 34 |
| Unemployed (Registered) (N = 166) | 0 | 32 | 0 | 1 | 28 | 1 | 1 | 37 |
| Other (N = 65) | 12 | 29 | 0 | 0 | 15 | 5 | 2 | 37 |
| Protestant | | | | | | | | |
| Full-time education (N = 713) | 26 | 41 | 0 | * | 5 | 2 | 1 | 26 |
| Full-time work (N = 310) | * | 61 | 0 | 0 | 7 | 1 | 2 | 29 |
| YTP (N = 313) | 1 | 46 | 0 | 0 | 19 | * | 3 | 31 |
| Unemployed (R) (N = 103) | 0 | 35 | 0 | 1 | 25 | 2 | 2 | 35 |
| Other (N = 39) | 8 | 33 | 0 | 0 | 13 | 3 | 3 | 41 |
| Catholic | | | | | | | | |
| Full-time education (N = 695) | 22 | 27 | 1 | 1 | 12 | 2 | 1 | 34 |
| Full-time work (N = 171) | 1 | 57 | 0 | 1 | 7 | 1 | 3 | 30 |
| YTP (N = 264) | * | 34 | 0 | 2 | 22 | 3 | * | 30 |
| Unemployed (R) (N = 63) | 0 | 27 | 0 | 2 | 32 | 0 | 0 | 40 |
| Other (N = 25) | 20 | 20 | 0 | 0 | 16 | 8 | 0 | 36 |

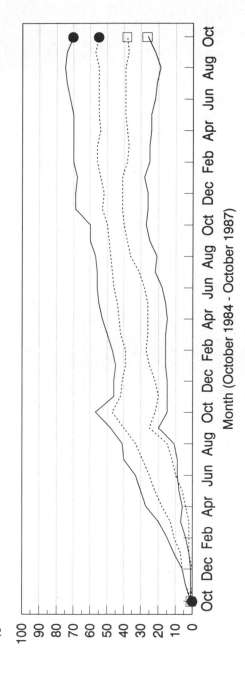

%

100
90
80
70
60
50
40
30
20
10
0

Oct Dec Feb Apr Jun Aug Oct Dec Feb Apr Jun Aug Oct

Month (October 1984 - October 1987)

Full-time work    Full-time work    Unemployed       Unemployed
(Protestant)      (Catholic)        (Protestant)     (Catholic)
●                 ●                 □                □

Note:

Protestant (N=190)
Catholic (N=143)

FIG. 5.4. Work and unemployment profiles of Protestants and Catholics who
were in YTP at age 16: Stage Two (weighted)

TABLE 5.4. *Most recent YTP scheme attended by main activity by religious affiliation at age 19* (%)

| YTP Scheme | All | | | | Protestant | | | | Catholic | | | |
|---|---|---|---|---|---|---|---|---|---|---|---|---|
| | Full-time work | Unemployed | Other | Base = 100% (N) | Full-time work | Unemployed | Other | Base = 100% (N) | Full-time work | Unemployed | Other | Base = 100% (N) |
| Workscheme | 73 | 22 | 9 | 275 | 72 | 22 | 6 | 160 | 67 | 20 | 13 | 109 |
| College-based | 65 | 33 | 3 | 80 | 70 | 28 | 2 | 54 | 54 | 42 | 4 | 26 |
| Training Centre | 63 | 34 | 3 | 111 | 62 | 36 | 2 | 61 | 64 | 32 | 4 | 50 |
| Community Workshops | 47 | 43 | 10 | 70 | 53 | 32 | 16 | 19 | 45 | 47 | 8 | 51 |
| Employer-based | (19) | (2) | (2) | 23 | (14) | 0 | (1) | 15 | (5) | (2) | (1) | 8 |
| Youthways | (8) | (8) | (3) | 19 | (3) | (4) | (1) | 8 | (5) | (4) | (2) | 11 |
| Other | (34) | (16) | (5) | 55 | (21) | (6) | (1) | 28 | (13) | (10) | (4) | 27 |
| N | 408 | 179 | 46 | 633 | 243 | 90 | 18 | 351 | 165 | 89 | 28 | 282 |

*Note*: Parentheses indicate cell sizes too small to calculate percentages.

home backgrounds than the Protestants (61%). Thus the different economic activity patterns described above could have been produced by the interrelationships which might exist between religious affiliation, YTP experience, and other relevant factors.

In order to examine more thoroughly the likelihood of the YTP participants being either in full-time work or unemployed, multivariate analyses were carried out. Log-linear analysis simultaneously examined the possible effects on the likelihood of being in a full-time job at age 19 (as opposed to being unemployed) of five variables: religious affiliation, gender, educational attainments, social background, and YTP experience, both as separate and as interacting variables.

The subsamples in the study which were selected as the most appropriate comparison groups in the analysis are as follows. Educational qualifications and social background were both dichotomized in order to eliminate empty cells in the analysis. Young people's educational performance (academic and vocational) was classified on the basis of whether they had obtained 5 or more O levels or their equivalent, while social background was divided on a 60:40 basis (60% from relatively less advantaged home backgrounds)—because it proved more feasible in terms of the distribution of the data.

The influence of religious affiliation, YTP experience, qualifications, social background, and gender on the weekly wages of the full-time employed at age 19 was examined for those who did not go straight from school to work using analysis of co-variance, with the length of time in employment as the co-variate. In this way it was hoped to take account of the varying lengths of time the young people had been in employment, due, for example, to their having continued in full-time education beyond age 16 or to having been in YTP.

All of the variables *except religious affiliation* had a significant impact on wage levels (see McWhirter *et al.*, forthcoming, for details). That is, Protestants and Catholics who were in employment for the same length of time earned approximately the same wages irrespective of their gender, social background, and qualifications and regardless of whether or not they had YTP training.

TABLE 5.5. *Results of log-linear analysis for effects of religious affiliation, sex, social background, educational performance, and YTP experience on economic activity of 19-year-olds who did not go directly from school to employment*

| Factors/levels | Observed proportions in full-time work (%) | F-ratio | Degrees of freedom | Significance |
|---|---|---|---|---|
| Qualifications | | | | |
| Five or more O levels | 75 | | | |
| Less than 5 O levels | 61 | 18.9 | 1 | * |
| Religion by social background by gender | | | | |
| Protestant | | | | |
| More advantaged social background (male) | 83 | | | |
| More advantaged social background (female) | 80 | | | |
| Less advantaged social background (male) | 64 | | | |
| Less advantaged social background (female) | 73 | | | |
| Catholic | | | | |
| More advantaged social background (male) | 58 | | | |

| | | | | |
|---|---|---|---|---|
| More advantaged social background (female) | 84 | | | |
| Less advantaged social background (male) | 58 | | | |
| Less advantaged social background (female) | 57 | 9.8 | 1 | * |
| Social background by YTP experience | | | | |
| More advantaged social background | | | | |
|   YTP experience | 83 | | | |
|   No YTP experience | 69 | | | |
| Less advantaged social background | | | | |
|   YTP experience | 64 | | | |
|   No YTP experience | 63 | 8.8 | 1 | * |
| Religious affiliation by YTP experience | | | | |
| YTP experience | | | | |
|   Protestant | 73 | | | |
|   Catholic | 65 | | | |
| No YTP experience | | | | |
|   Protestant | 75 | | | |
|   Catholic | 56 | 5.2 | 1 | † |

* $p < 0.01$.

† $p < 0.05$.

The results of the log-linear analysis (Table 5.5), indicate that all of the five explanatory variables considered made a significant contribution, in some way or another, to determining an individual's chances of being in a job at age 19. The most important finding in the present context is that YTP interacted significantly with social background and religion (but not qualifications or gender). Religious affiliation also interacted significantly with social background and gender. Finally, only one variable, qualifications, was found to be equally influential across all the different subgroups in the analysis. For the sake of brevity, only the significant results involving religious affiliation will be reported in detail. (See McWhirter *et al.*, forthcoming, for an analysis of the significant findings relating to gender.)

As may be seen in Figure 5.5, YTP experience was of value in obtaining full-time employment at age 19 for those Catholics who did not go straight from school to work. Protestant YTP participants did not experience the same market advantage compared to their co-religionists, but they were still more likely than the Catholic YTP participants to find a job. This important finding, which suggests that YTP has helped to narrow the gap in the employment rates of Protestants and Catholics, is

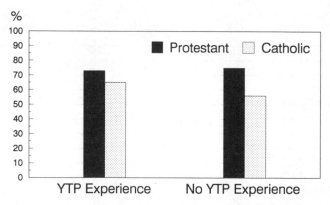

FIG. 5.5. Proportion of economically active in full-time work by religious affiliation and by YTP experience. Economically active includes those in full-time work or registered unemployed but excludes those who went directly from school to work

independent of whatever other variations might occur in employment rates due to social background, educational performance, or gender.

YTP experience did, therefore, contribute to some extent in reconciling the different employment rates between the two religious groupings. In general, of those employed at age 19 who had YTP experience, more of the Workscheme participants than those who had experience of the full-time training schemes only were satisfied with the help received from the Programme in finding employment. In both instances, and in accordance with the log-linear findings, it is notable that more Catholics than Protestants were content with the assistance that YTP training and Workscheme had provided in obtaining employment (YTP training—Protestants 53%, Catholics 59%; Workscheme—Protestants 61%, Catholics 74%).

While the information is not complete in this regard, there is some evidence to suggest that young Catholics and Protestants were also obtaining different types of jobs after completion of their YTP training. Due to the varying proportions of the different groups of employees falling into the missing/unclassified category (Table 5.6) caution must be exercised in comparing the different types of jobs obtained by the groups. The data suggest that YTP-trained workers, particularly the Protestants, were more likely than the non-YTP groups to have skilled manual jobs at age 19 and less likely to be in intermediate and junior non-manual employment. The data also suggest that YTP-trained Protestants and Catholics were more likely than their non-YTP counterparts to be working in the construction and manufacturing sectors and less likely to be in the public sector. Irrespective of YTP experience, a greater proportion of Protestant workers than Catholics were in private services and manufacturing.

Finally, the results suggest that there is a complex relationship in the job search during the first few years after entering the labour market between religious affiliation, gender, and social background for those young people who do not find a full-time job immediately upon leaving school (Figure 5.6). At the risk of over-simplifying the results, a number of trends are apparent which, nevertheless, contain notable exceptions.

TABLE 5.6. *Socio-economic group and industrial distribution of full-time job at age 19 for those who did not go directly from school to work: by YTP experience by religious affiliation* (%)

| | YTP experience | | No YTP experience | |
|---|---|---|---|---|
| | Protestant | Catholic | Protestant | Catholic |
| A. Socio-economic group | | | | |
| Employers and managers | 1 | 0 | 2 | 1 |
| Intermediate non-manual | 3 | 1 | 7 | 4 |
| Junior non-manual | 21 | 20 | 29 | 26 |
| Personal services | 5 | 7 | 4 | 9 |
| Skilled manual | 35 | 26 | 13 | 13 |
| Semi-skilled manual | 13 | 14 | 16 | 15 |
| Unskilled manual | 3 | 4 | 2 | 2 |
| Other | 2 | 2 | 8 | 3 |
| Missing | 17 | 25 | 20 | 27 |
| TOTAL | 100 | 100 | 100 | 100 |
| B. Industrial distribution | | | | |
| Manufacturing | 27 | 21 | 21 | 17 |
| Construction | 14 | 16 | 3 | 4 |
| Private services | 35 | 30 | 36 | 32 |
| Public services | 6 | 6 | 16 | 17 |
| Other | 18 | 26 | 24 | 30 |
| TOTAL | 100 | 100 | 100 | 100 |
| N | 337 | 258 | 186 | 179 |

*Note*: Totals may not add to 100 due to rounding.

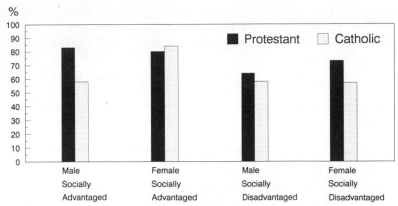

FIG. 5.6. Proportion of economically active in full-time work by religious affiliation and by sex by social background. Economically active includes those in full-time work or registered unemployed but excludes those who went directly from school to work

Catholics, in general, fared less well than Protestants in obtaining jobs (apart from females from more advantaged home backgrounds). Young people from more advantaged home backgrounds were more likely to be employed than those from relatively less advantaged backgrounds (apart from Catholic males). Given that the impact of YTP experience applies irrespective of the gender and home background of Protestants and Catholics, the types of jobs obtained by the latter groups will not be examined here (see McWhirter *et al.*, 1989).

The study findings, although complex at times, indicate the value of examining work-related data in detail using multivariate analyses: it would be misleading not to acknowledge that the process of moving from school to work is complicated, multifaceted, and dynamic and that young people manifest a variety of characteristics at the same time. In this context it is important to note that the log-linear analyses by necessity included a limited number of variables. The goodness of fit obtained in the model was, however, sufficiently high to enable us to conclude that other possibly relevant factors, as they are currently available in the database, would not have contributed markedly to the results. This is not to say that variables such as the choice of subjects at school and area of residence (related as it is to the structure of employment opportunities in different local labour markets) are unimportant in the job hunt, but they did not form part of this study (see Cormack and Osborne, 1983).

Educational profiles for Catholics and Protestants are known to differ with regard to the type of subject qualification obtained. Recent information (Osborne and Cormack (1989*b*)) confirms the greater likelihood of Catholics qualifying in arts-related disciplines while Protestants continue to favour subjects with a greater science content. Additional coding of the cohort database should enable due account to be taken of the obvious implications such information has both for employment rates and the types of job obtained by the two religious groups in the study.

## CONCLUSIONS

The available evidence in the study suggests that Protestants and Catholics appear to have equal access and equity of treatment in the Programme as a whole, but clearly a more detailed examination of the different types of schemes would be necessary in order to come to a firm conclusion. Whilst the new PPRU longitudinal study of YTP participants should be of assistance in this regard, the present findings revealed imbalances in the uptake of different types of provision by the two groups (e.g. more Protestants in college-based training schemes and in Workscheme and more Catholics in Community Workshops). Also, earlier evidence (Whyte *et al.*, 1985) indicated that religious segregation on the training schemes did occur.

The issue of the integration of the two groups is of less importance to the employment debate in Northern Ireland than it is to the issue of community relations (Whyte *et al.*, 1985), provided that the different types of schemes are equally available throughout different areas of the Province. As we have noted, the post-YTP employment rates vary considerably between the different types of scheme.

The study also showed that overall there was equal uptake of YTP by Protestants and Catholics but uncovered different factors which led to this situation. A greater proportion of Catholics remained in full-time education beyond school-leaving age and more of the Protestant labour market entrants secured full-time employment immediately after leaving school. Thus, despite equal participation rates for YTP, the two groups did not have an equal 'need' to join the Programme.

In terms of equity of treatment and equality of outcome it is notable that, although fewer of the Catholics than the Protestants thought highly of YTP when they were in the fifth form, the two groups who had experience of the Programme held equally positive views about it afterwards.

In the broader context of the transition from school to work of unemployed school leavers the study provided some tentative answers regarding the role of religious affiliation and YTP experience in this process. It is of compelling interest to note that the provision of employment training, in general, contributed to reducing the disparity in employment rates between the two religious groups—although this effect is in many ways outside the control of the providers of YTP. While still less likely than Protestant trainees to be employed, Catholics who were unable to go directly from school to work stood a better chance of obtaining a job than if they had opted not to join the Programme. As indicated earlier, it is important to note that this is only one of a number of different factors contributing towards the variations in employment rates between the two religious groups.

There appears to be a considerable way to go before equal opportunities for young Catholics in employment in the wider labour market will be achieved. However, despite the fact that the prime purpose of the Youth Training Programme is not to act as a compensatory employment measure for particular disadvantaged groups in the Province, the results suggest that the Programme has made a positive impact on the employment prospects of Catholic labour market entrants who did not move directly from school to work.

# 6

## Equality of Opportunity in the Workplace: A Survey of Northern Irish Employers

*Gerald Chambers*

### INTRODUCTION

Until the research reported on in this chapter there had been no systematic attempt to assess the impact on employers of recent fair employment legislation in reducing discrimination and promoting equality of employment opportunity between the two religious communities in Northern Ireland. Allegations of discrimination in employment have been a long-standing grievance of the Catholic community in Northern Ireland (Cameron Commission, 1969). Although in the past there has been some debate about the true extent of that discrimination and about its causes and origins (Whyte, 1983), there can be little doubt that Catholics and Protestants in Northern Ireland have quite different occupational profiles, or that there is differential access to employment opportunities, or that such differences contribute to inequalities in levels of affluence and standard of living. These differences are nowhere more apparent than in a current Catholic male unemployment rate which is two and a half times greater than the Protestant equivalent (Smith, 1987*a*).

While Northern Irish fair employment legislation has its origins in a committee of inquiry set up by the British government after the imposition of direct rule, and while the legislation, the 1976 Fair Employment Act, that followed evolved directly from that committee's recommendations (MHSS, 1973), it is also the case that the Northern Irish fair employment legislation had many resemblances to United Kingdom race and gender legislation, although there are important differences in its administration. In common with gender and race legislation, there is a declarative prohibition on discrimination, and channels of redress are established for individuals who feel they have been the victims of discrimination. Secondly, there is a statutory agency with powers to investigate employment practices and to provide advice and guidance to employers on the promotion of equality of opportunity. Thirdly, the legislation provides for a voluntary code of employment practice which gives advice and guidance to employers on what steps they should take to avoid discriminatory practices and promote equality of opportunity.

The Northern Irish legislation has recently been the subject of a major review by the Standing Advisory Commission on Human Rights, which drew on the research reported below (SACHR, 1987). The main aim of the research was to find out about current policies and practices in employment. In particular, we wanted to gather information about recruitment and promotion practices and to find out about the prevalence of a range of equal opportunity measures. It was our aim to collect empirical information as a way of addressing and providing answers to the following questions: how do employers currently go about recruiting labour, and are current recruitment practices contributing to inequalities in employment? How active have employers been in adopting measures designed to promote equality of opportunity and afford open access to jobs? Is the present regulatory and legislative framework adequate to achieve the objectives of employment equality and, if not, in what way might it be improved?

In order to seek answers to these questions a survey of approximately 250 workplaces in Northern Ireland was undertaken during the autumn of 1986. The respondents were personnel managers where such a role existed, or members of the senior managerial staff. Small, medium, and large establishments were surveyed and the sample was designed so as to be representative of employing establishments throughout Northern Ireland. Survey work was supplemented by detailed case-studies of selected establishments.

This chapter, then, does not try to establish how much, or how little, discrimination in employment there is in Northern

Ireland. Rather, the subject of discrimination is approached from another angle: an assessment of the extent to which equality of opportunity in employment, and thus the conditions which disallow job discrimination, has become a reality. Three issues which shed light on this question are considered: first, have employers in Northern Ireland implemented any measures designed to promote equality of opportunity? In particular we are concerned here with those measures which featured in the government's first *Guide to Manpower Policy and Practice* (DMS, 1978). Secondly, what are the recruitment practices and job appointments procedures of Northern Ireland's employers and what impact will such procedures have on creating job equality? Finally, what information do we have about the impact of the regulatory work of the Fair Employment Agency (FEA) in influencing workplace practices?

## PROMOTING EQUALITY OF OPPORTUNITY

Before enquiring into the prevalence of measures specifically intended to promote equality of opportunity, we asked a number of questions designed to assess the organization's awareness of the framework of equality legislation and to establish how the legislation impinged on its operations.

First, when we asked respondents a general question about the impact of the Fair Employment Act on their policies and practices we received the following responses: the overwhelming majority of respondents (85%) reported that the Act had had little or no impact on personnel policies at all, while the remainder were more likely to report that the Act had some impact (12% of the total) than a great deal of impact (3%).

When asked to describe the ways in which the Act had had an impact the small group of respondents for whom this was a relevant question gave responses which indicated that there had been changes in the *awareness* within the organization about the issue of discrimination, and only a few responses were concerned with real changes in practices or procedures. So even among those who said that the Act had had an impact, only a few were able to give examples of practical consequences.

Secondly, we asked about knowledge of, and action taken in

relation to, the *Guide to Manpower Policy and Practice.* We found that nearly two-thirds of employers were not even aware of the existence of the *Guide,* although public-sector establishments had a higher level of awareness than those in the private sector. Thus, the cautious recommendations to be found in that document had not by 1986 been successfully communicated to the majority of employers.

Thirdly, we thought it would be informative to enquire into whether the establishment had the objective of achieving a balanced workforce. We asked the following question: 'Would you say that it was the policy of this establishment to set about to achieve a religiously balanced workforce, or is it policy to let things find their own level?' More than nine out of ten respondents replied that it was the organization's policy to let things find their own level. There was a surprising degree of un-animity on this question in all industrial sectors and the only group of respondents who showed any degree of variation in their responses were large employers, but even in this group eight out of ten said they let things find their own level.

Such a response indicates that employers are not in general predisposed towards the attainment of a religious balance as a priority, or because such a balance is in itself a good thing. Clearly if a policy of 'letting things find their own level' is combined with an absence of measures designed to ensure that the establishment attracts applicants from both communities, then the result can only be the perpetuation of inequalities and imbalances. What is perhaps surprising is that so few employers saw the attainment of balance in the workforce as an objective in itself.

Nevertheless we believe that these responses understate the extent to which some employers go out of their way to ensure that they have a sufficient number of employees from each community, so that work which requires servicing or carrying out activities in both communities can continue to be done. Some employers are reluctant to admit that these considerations exist because they are unsure of the legality of operating in this way and are therefore reticent about admitting to practices which could be seen as illegal. So while attainment of balance may not be seen as a stated goal, there are numerous examples of employers who seek to construct a mixed workforce because the nature of their activities requires employees from both communities.

For example, the manager of a large retail unit in a busy shopping area had decided that the religious balance of the workforce (over fifty staff in total) had come to be unrepresentative of the religious balance of the surrounding area. An opportunity for expansion had given the company the chance to rectify this by increasing the proportion of Catholics in the workforce through a policy which the manager described as 'positive discrimination'. Once the correct balance had been achieved an employee who left was replaced by a person of the same religion so that the religious balance was maintained. The manager was of the view that his policy made good business sense and could be justified on purely commercial grounds: in order for his business to thrive it needed to maintain customer support. His policy of 'positive discrimination' in order to rectify an imbalance helped to re-establish the credibility of the company in the eyes of the local community.

Other employers find themselves providing services on a cross-community basis: for example, a local authority thought it necessary to ensure that its leisure centre facilities were not monopolized by only one community and sought to employ leisure centre staff who were both Catholic and Protestant. Thus when the authority appointed a Protestant as manager of the centre there was pressure to appoint a Catholic as assistant manager. In this way religion became an inevitable consideration in recruitment. Similarly, a transport business was involved in collecting and delivering in both Catholic and Protestant areas. Not only was there a need to recruit both Catholics and Protestants as drivers but there was in some localities a need to ensure, for security reasons, that there was a match between the religious affiliation of the drivers and the areas they had to operate in. In these ways the religion of a job applicant is inevitably taken into account.

Most people in Northern Ireland probably understand equal opportunities to mean people from both communities competing equally for the same sorts of jobs. The visible outcome of such a policy would be that both Catholics and Protestants would be working together for the same employers under the same terms and conditions. The irony of the situation is that employers who are seeking to create a mixed workforce in the ways described above were probably acting illegally because the 1976 law did not distinguish between actions which are designed to include someone because of his or her religion and actions which are designed to exclude.

The apparent lack of awareness and attention to equal opportunity strategies, objectives, or goals has therefore to be seen in context and with qualification. First, it is clear that equal opportunities goals, even ones which may cross the boundaries of legality, are not considered by all employers to be incompatible with the pursuit of commercial goals. Secondly, some employers will adopt a pragmatic recruitment strategy in order to maintain cross-community services. Thirdly, there is an increasing awareness among employers of the need to be able to defend their practices if challenged by outsiders or by the FEA to do so. Some employers will therefore have a token minority presence in the hope that such a strategy will ward off difficult questions and the attention of the FEA.

## EQUAL OPPORTUNITY MEASURES

The findings of the research clearly show that more than ten years after the introduction of the 1976 fair employment legislation few employers have taken practical steps to implement equal opportunity measures. Table 6.1 lists eight equal opportunity measures which we examined in the course of our research and indicates the frequency with which these were implemented in our sample of workplaces. It can be seen that the most frequently implemented measure was signing the Declaration of Principle and Intent (62% of establishments) and that this was the only measure to be implemented by a majority of establishments.

It should be remembered that employers who signed the Declaration were not thereby committing themselves to any further action. Nor was there any scrutiny of their policies and practices after signing. The high rate of signing is thus

TABLE 6.1. *Proportion of workplaces adopting eight equal opportunity measures* (%)

| Equal opportunity measure | Total | Size of establishment in terms of number of employees | | | |
|---|---|---|---|---|---|
| | | <10 | 10–49 | 50–99 | ≥100 |
| Signed declaration | 62 | 38 | 63 | 81 | 87 |
| Reviewed recruitment procedures | 21 | 2 | 22 | 33 | 47 |
| Initiated contact with schools | 20 | 12 | 20 | 12 | 43 |
| Set up disciplinary procedures | 10 | 1 | 7 | 7 | 37 |
| Operates transport system | 9 | 4 | 8 | 14 | 13 |
| Sought advice of the FEA | 8 | 9 | 4 | 17 | 17 |
| Has a written statement or policy | 8 | 1 | 4 | 4 | 41 |
| Has a monitoring system | 5 | 0 | 5 | 9 | 11 |
| Number of workplaces | 243 | 30 | 80 | 59 | 74 |

*Source*: Chambers, 1987.

explained by the fact that it is a paper policy only. Even employers who had a bad record in relation to equal opportunities were still able to sign the Declaration in order to gain access to contracts and grants in the knowledge that it was unlikely that there would be any scrutiny of their policies by the FEA. These facts help to explain employers' reasons for signing the Declaration. We asked respondents what benefits they thought might have accrued to them as a result of having signed. The most important finding was that nearly half of all respondents were unable to list any benefits at all. However, the largest single benefit, mentioned by just over a quarter of all signatories, was that signing allowed access to contracts and grants.

Despite the fact that signing the Declaration was a paper policy only, there were indications that signatories had more extensive equal opportunities policies than those who had not signed. Workplaces which were not signatories of the Declaration were less likely than others to have implemented other equal opportunity measures. For example, 23% of signatories had also reviewed their recruitment practices in the past three years compared with only 5% of those who had not signed; 12% of signatories had set up disciplinary procedures compared with only 1% of non-signatories. In all categories, signatories came off better than non-signatories.

However, measures which required employers to change practices or procedures had a much lower prevalence. The 1978 *Guide to Manpower Policy and Practice* stated:

It is to be expected that practices which, whether intentionally or not, tend to favour one section of the community at the expense of others, will be reflected in the religious make-up of the employees or membership of the organization concerned (the pattern of membership). The monitoring of patterns and trends of employment or membership can therefore be a useful means of ascertaining whether or not equality of opportunity is provided. Employers and vocational organizations are therefore recommended to consider identifying the patterns and trends within their organizations. (DMS, 1978: 15–16)

Yet very few workplaces (only 15%) had by 1986 adopted a formal system for identifying 'the pattern of employment', to make use of the euphemism coined by the *Guide*. This figure

increased to almost 10% for medium-sized and large employers. Monitoring requires the collection of information on
religious affiliation in a systematic way. In Northern Ireland
this can be done by asking employees or job applicants to state
religious affiliation, or by collecting information on schools
attended. Our findings show that while only 4% of establishments kept a record of their employees' religious affiliation on
file, a much higher proportion, about 40%, kept information on
schools attended.

However, most managers have a very clear idea of the
religious affiliation of members of the workforce which is quite
independent of any formal record-keeping. As part of our
research we collected information on the religious composition
of all the workforces we researched. Although we had been
warned that we would experience great difficulty in collecting
this information, we were successful in 90% of cases in finding
out whether the workforce had a Catholic or a Protestant
majority. Furthermore, we were successful in eight out of ten
cases in finding out the proportion of Catholics and Protestants
in the workforce to the nearest 5% based on estimates provided
by the respondent manager.

Clearly this knowledge of workforce composition was the
result neither of systematic record-keeping nor of monitoring,
since only a minority of establishments admitted to such
practices. When we asked respondents on what basis they had
been able to provide us with a compositional analysis of
religious affiliation, by far the most frequently cited reason was
'personal knowledge of most of the workforce'. Our assessment
is that assigning a religious affiliation to employees poses no
major difficulties nor does it conflict with the way people
generally think about and evaluate each other's status and
social position in everyday life. Opposition to monitoring is,
however, widespread among Northern Irish employers and
this can probably be explained in three ways. First, many
people believe that merely acknowledging the existence of the
sectarian divide gives legitimacy to it; monitoring clearly con-
fronts this view head on since the principle behind it is that the
only way to overcome the sectarian divide is by first of all
making it explicit. Secondly, there is a fear that collecting

information on religious affiliation will leave the employer open to charges of discrimination. Probably not much credence should be given to this concern since those who are minded to discriminate are more likely to do so on the basis of informal knowledge derived from name, manner of speech, and area of residence. Thirdly, the greatest opposition to monitoring comes from those employers who are concerned about the imbalances that a monitoring exercise would uncover.

Turning now to other equal opportunity measures, it was apparent that prevalence rates for these remained low. For example, we looked at the proportion of establishments which had recently examined their recruitment and promotion methods. It can be argued that an essential aspect of the promotion of equality of opportunity is that employers should be conducting regular appraisals or reviews of the effectiveness of their practices and procedures, as is now part of the new legislation. We found that reviews had been conducted at 21% of workplaces within the past three years but that the likelihood of a review increased with the size of the workplace.

What did reviews of recruitment procedures amount to and to what extent were these prompted by equal opportunities concerns? Any establishment with a personnel department, even if it consists of only one officer, is likely to be engaged in reflection about the extent to which its procedures are helping it to achieve organizational goals. It is likely that most of the 'yes' responses to whether or not reviews had taken place related to reviews of this general type. Certainly it was rarely the case that reviews were specifically conducted with religious equality of opportunity in mind.

Nine per cent of workplaces provided transport for their employees. If an establishment provides transport in order to bring its employees to work it need not necessarily be with the intention of promoting equality of opportunity. Few would question, however, that the accessibility of the workplace is an important factor in determining the composition of the work-force, or that in Northern Ireland the location of a workplace can either inhibit or enable the promotion of equality of opportunity. Transport can be provided so as to overcome the drawbacks of location.

There was no evidence that any of these transport schemes had been drawn up primarily as a 'busing' measure. However, it was certain that the withdrawal of company transport could have implications for equal opportunities. One illustration of this comes from an employer whose premises were situated adjacent to a strongly loyalist housing estate in mid-Ulster and who provided buses in order to bring workers in for the evening and late shifts and to take them home again. The company's buses went into both Protestant and Catholic housing estates in the surrounding area. If the transport system had not existed it is likely that many of the company's present employees on night shifts would no longer have been able to get to work. Instead the company would have had to rely on a more local labour market which was exclusively Protestant, with consequent repercussions for the composition of the work-force at the plant.

If an organization is one which takes the issue of equality of opportunity seriously it is likely that it will have developed a written statement which encapsulates its policy and which could in some form or another be communicated to its employees. (Such a course of action was recommended in the 1978 *Guide to Manpower Policy and Practice*.) Yet only 8% of establishments had such statements and the vast majority of these were among the larger employers, where 41% of establishments employing more than 100 people claimed to have such a statement. Respondents who reported that they had such a statement were asked how it was communicated to their employees. The most frequently used method of making it known was to have a copy in a staff handbook (which not all workplaces necessarily make available to each employee). In only a very few instances was a copy of the statement given to each employee individually, or posted on a noticeboard.

Related to having a policy statement is the need to have procedures for dealing with allegations of discrimination. Again only 10% of all workplaces but more than one-third of large workplaces had developed special procedures. Even when formal procedures existed for taking up complaints of discrimination serious difficulties stood in the way of dealing with complaints, and employees could find that trade unions were unable or reluctant to become involved or to give practical assistance. Few shop stewards have experience of taking up discrimination complaints in a formal way and there is extreme reluctance to get involved in what are seen as sectarian issues. In general, when a shop steward says that he or she is there to represent all union members what is meant is that issues which draw attention to the sectarian divide, but which might be crucial to *some* members, cannot be taken up. Employees will often find that they are left to their own resources if they wish to make a complaint or pursue a grievance.

It is a useful measure of how much importance employers attach to the issue of fair employment to examine how often they themselves initiated contact with the FEA in order to seek advice and guidance. We therefore asked respondents how often they had 'sought general advice from the FEA about fair employment issues'. However, only 8% of respondents reported voluntary contact with the FEA in this way and it is therefore clear that the FEA is not being used as a resource agency to any great extent by employers in Northern Ireland.

Finally, one indicator of the extent to which an establishment can be thought of as an equal opportunity employer is whether or not it seeks to forge links with the community through contacts with local schools and educational establishments. We have examined contacts initiated by employers themselves, as opposed to approaches made by schools to employers. One-fifth (21%) of establishments could be described as having initiated contacts with schools. We are not able to determine if an establishment had links with both maintained and controlled schools as this would have required an inventory of local schools and an attempt to measure frequency of contact in each sector. This was beyond the scope of a survey interview. (However, such an inventory has now been compiled by the government as an aid to monitoring (DED, 1987*a*).) Obviously, links made exclusively with schools in one sector only would be detrimental to equality of opportunity. It is, therefore, important to recognize that a greater degree of contact cannot in any simple way be equated with greater provision of equal opportunity and indeed that if all contacts were with one community only then the very reverse would be the case.

To summarize, Table 6.2 shows the number of equality

TABLE 6.2. *The number of equal opportunity measures adopted by workplaces (%)*

| Number of measures | Total | Size of establishment in terms of number of employees | | | |
| | | <10 | 10–49 | 50–99 | ≥100 |
| --- | --- | --- | --- | --- | --- |
| 6–8 measures | 1 | 0 | 0 | 0 | 12 |
| 3–5 measures | 17 | 6 | 15 | 23 | 45 |
| 1–2 measures | 55 | 37 | 63 | 72 | 39 |
| None | 26 | 57 | 22 | 5 | 4 |
| Mean | 1.4 | 0.7 | 1.3 | 1.8 | 3.0 |

*Note*: Most workplaces in the public sector do not make local decisions about the adoption of policies and procedures and in the above table these workplaces have been excluded. Hence the total number of workplaces is reduced to 206 for this analysis.

*Source*: Chambers, 1987.

measures implemented in workplaces. The results show that only 1% of workplaces had implemented all eight measures and that these were in the group of large employers. A sizeable proportion (26%) had implemented no equal opportunities measures at all. The mean number of measures to be implemented by all workplaces was only 1.4, and for large workplaces the mean was 3.

We can sum up what has been said so far in the following way: it was the view of our respondents that the Fair Employment Act and, to a lesser degree, the *Guide to Manpower Policy and Practice* had had little direct impact on employment practices. Evidence from the survey indicated that only a few employers went out of their way to achieve a religiously balanced workforce; but there was also evidence that some employers sought to achieve a degree of mix in the work-force, for pragmatic reasons, such as the need to ensure that cross-community functions were maintained.

Very few establishments were monitoring the religious composition of the work-force. There was widespread hostility towards monitoring and most employers did not understand its relationship to the promotion of equality of opportunity. Establishments could quite happily describe themselves as equal opportunity employers but be completely opposed to monitoring. Even some employers who had been investigated by the FEA remained totally opposed to monitoring. Few establishments systematically collected information on religious affiliation but a greater number collected information on schools attended. Yet managers at all but the very largest of establishments knew about the religious affiliation of members of the work-force. This reluctance to organize in a systematic way information that was common knowledge is explained in large part by anxieties about what monitoring might uncover. Monitoring is by itself not a solution to work-force imbalances and will only be effective as part of a broader affirmative action programme. Most of the component parts of such a programme were not being pursued by employers.

## RECRUITMENT AND APPOINTMENT PROCEDURES

Many factors influence an organization's institutional arrangements for securing, retaining, and shedding labour, including for example the general availability of labour and levels of unemployment, the competitiveness of the organization's wage rates in comparison to other organizations in the local economy, and the organization's dependence on new technology and advanced methods of production. In consequence, those responsible for recruiting labour work within constraints set by production and market-oriented imperatives.

A useful distinction has been made between, on the one hand, organizations which seek to create and sustain an internal labour market by restricting the points of entry to one occupation, or to the bottom rung of the ladder, and, on the other hand, organizations which are more open to penetration at various occupational and status levels (Ford *et al.*, 1984). In this conception, organizations which primarily depend on such an internal labour market will fill vacancies through in-house advertising and promotion, or by internal transfer from one part of the organization to another.

It has been argued that the creation of internal labour markets provides benefits for both management and workers: for management, reliance on internal labour markets can facilitate the creation of a stable, reliable, and hence co-operative workforce, while, by offering employees a career or a future within the organization, protection is offered from the insecurity of the job markets and existing employees are protected from external competition from outsiders.

No labour force is wholly closed or can rely indefinitely on supplying all its labour requirements from the internal market. One method by which an organization can draw in new blood but at the same time retain control over the quality of the intake, thereby minimizing the potential for disruption to organizational goals or production targets, is by using current employees as recruiters of new labour. Such institutional arrangements have become known as the extended internal labour market (Manwaring, 1984).

The precise nature of the institutional arrangements for

recruitment are not, then, purely a function of personnel policies or the expression of an overriding need to find at any cost the best person to perform a particular task or role. A variety of factors determine and help to sustain the organization's recruitment channels: pressure to restrict the points of entry and to satisfy labour demands through an extended internal labour market may be maintained by the collective strength of craft unions, by the inability of the organization to support competitive recruitment mechanisms, or by the perceived need to induce an organizational *esprit de corps*. On the other hand, product innovation, technological change in production processes, and local labour shortages are some of the factors which are likely to lead to a broadening of the recruitment base and the undermining of reliance on internal markets.

What linkage is there between this typology of labour markets and equality of opportunity? There can be no simple equation between the promotion of equality of opportunity and exclusive reliance on external labour markets for recruitment purposes. Reliance on external labour markets will not broaden the gender, race, or religious profiles of those recruited unless comprehensive equality procedures are instituted at all points of entry. Moreover, a greater degree of equality of opportunity may be afforded by reliance on internal markets and by restricting entry to the lowest rung of the ladder. This is because an organization which permits entry only at the lowest rung of the ladder (in terms of, say, age, skill level, and occupational status) may avoid taking on board the social inequalities inherent in the external pool of labour. By confining intake to the bottom of the rung and only to those with minimum skills, training, or qualifications, a more broadly based and representative labour force may be attained. This option is, however, best suited to an organization where tacit skills which can only be learned in the course of doing the job are more valuable than formal skills learned through an apprenticeship.

However, if equality of opportunity is a goal, then organizations which have limited points of entry, and which fill vacancies primarily by career progression and internal promotion will have to pay more attention than organizations which are penetrated by external markets at all levels to the removal of requirements and criteria which might be creating career blockages for groups of employees. Organizations relying on internal labour markets may have to put more effort into the development of objective and standardized criteria for deciding on eligibility for progression and into the development of formalized promotion procedures.

The extended internal labour market probably poses the greatest obstacle to the promotion of equality of opportunity since reliance on current employees to recruit new labour is likely to reproduce existing imbalances and inequalities. Moreover, it is unlikely that the extended internal labour market offers nomination powers equally to all employees and such powers may be restricted to those, such as shop stewards, or employee representatives, who have greater access to the personnel function.

In the course of our survey of employers we have gathered a great deal of information about recruitment and appointments procedures and this data enables us to comment on the implications of current procedures for the extension and promotion of equality of opportunity. We asked all employers about vacancy rates in the past year in order to get some idea of the opportunities organizations had to put their recruitment methods into action. We asked only about permanent vacancies and not for information about temporary or casual work that came up. We found that just over 70% of establishments had had experience of recruitment in the previous twelve months and that a majority of establishments in all industrial sectors but one had taken on new workers. Vacancies were not, however, occurring equally among all categories of employee: they occurred most often for non-manual employees and least often for skilled workers.

If there had been recruitment in the previous twelve months, respondents were asked about the methods used to find people to fill the vacancies that had occurred during that period. They were shown a list of fifteen possible methods of recruitment and asked to list all the methods used to recruit new workers. Establishments invariably used more than one method for recruiting. The average number of methods used were 2.3, 3.2, and 2.2 for skilled, unskilled and semi-skilled, and non-manual respectively.

As a result of asking the question in this way we were able to gather information from each organization about the range of methods used over a period of one year, but this meant that we did not necessarily gather data about the method of recruitment for the successful appointee. It could have been the case, for example, that an employer regularly placed vacancies with the job centre but never actually appointed candidates who were sent to him/her in this way.

Table 6.3 shows the frequency with which a variety of recruitment methods were used for three categories of worker. For skilled manual workers we can see that the five methods most frequently used by our group of Northern Irish employers were, in order of frequency of use, notifying the job centre, advertising in Northern Irish dailies, advertising in local weeklies, waiting lists, and personal contacts. For unskilled and semi-skilled manual workers the most frequently used methods were notifying the job centre, personal contacts, advertising in local weeklies, waiting lists, and advertising in Northern Irish dailies. For non-manual staff, advertising in Northern Irish dailies was by far the most frequently used method, followed by notifying the job centre, advertising in local weeklies, personal contacts, and waiting lists.

It can be seen from this list of the top five methods that informal methods come into play for all categories of workers. Two informal methods of recruitment in particular were used a great deal: relying on the personal contacts of existing employees to find new workers, and hiring from a waiting list of interested people. These informal methods feature more prominently for unskilled and semi-skilled workers than for the other two groups.

Other interesting points to note are that 'contacting ex-employees' is quite a commonly used method for skilled workers but is less frequently used for unskilled and almost never used for non-manual staff. As might perhaps be expected, private employment agencies and advertising in specialist journals or magazines are used in recruiting non-manual workers but not to any notable extent for other categories

TABLE 6.3. *Recruitment methods for employees recruited in the previous twelve months* (%)

| Recruitment method | Skilled manual | Unskilled and semi-skilled | Non-manual |
|---|---|---|---|
| Advertise in GB daily newspapers | 4 | 0 | 6 |
| Advertise in NI daily newspapers | 33 | 23 | 53 |
| Advertise in local weekly newspapers | 38 | 28 | 27 |
| Advertise in journals or magazines | 0 | 0 | 9 |
| Through personal contacts or existing staff | 22 | 32 | 18 |
| Through casual enquiry | 2 | 7 | 2 |
| Offer jobs to temporary or casual employees | 5 | 9 | 3 |
| From waiting list | 27 | 23 | 13 |
| Notify Job Centre/PER/Careers office or local employment office | 41 | 49 | 41 |
| Contact ex-employees | 17 | 4 | 1 |
| Contact private employment agency | 1 | 3 | 12 |
| Let trade union nominate or provide people | 1 | 1 | 0 |
| Notify head office for internal transfer | 1 | 1 | 4 |
| Through apprenticeship schemes | 9 | 7 | 2 |
| Through youth training or community programme schemes | 16 | 21 | 9 |
| Number of workplaces | 79 | 127 | 150 |

*Source:* Chambers, 1987.

of worker. Employers frequently use the Youth Training Programme for the recruitment of permanent unskilled and semi-skilled manual workers (see Chapter 5).

In order to present this wide range of different recruitment methods in a more accessible format we have grouped them under four different headings according to the degree of formality of the procedures used. Thus we have devised four types of recruitment path: advertising, other formal methods (notifying the job centre or private employment agency), semi-formal methods (internal transfer, apprenticeship schemes, or YTP etc. schemes), and informal methods (personal contacts, taking on temporary workers, waiting lists, offering jobs to casual enquirers, contacting ex-employees, or accepting nominations from the trade union).

Focusing on skilled manual workers first, we can see from Table 6.4 that advertising is the most commonly used of the four paths, followed closely by informal methods and by other formal methods. For unskilled and semi-skilled manual workers other formal methods are in first place, followed closely by informal methods and then by advertising. For non-manual workers advertising is in first place followed by other formal methods, with informal methods being a long way behind and a

TABLE 6.4. *Summary of recruitment methods used in the previous twelve months* (%)

| Recruitment methods | Skilled manual | Unskilled and semi-skilled | Non-manual |
|---|---|---|---|
| No recruitment | 46 | 34 | 31 |
| Advertise | 31 | 26 | 43 |
| Other formal | 22 | 34 | 35 |
| Semi-formal | 14 | 17 | 10 |
| Informal | 27 | 30 | 19 |
| Other methods | 7 | 14 | 16 |
| Number of workplaces | 143 | 178 | 198 |

*Source*: Chambers, 1987.

much smaller percentage using semi-formal methods. As a consequence of presenting the data in this way the important part played by informal methods in the recruitment of both categories of manual worker can be seen more clearly. Informal methods do play a role, but a much less significant one, in the recruitment of non-manual staff.

To what extent did recruitment methods vary with the size of the establishment and was there any evidence that larger establishments were likely to adopt formal recruitment procedures more frequently than smaller ones? Our findings show that large establishments have generally used a greater variety of recruitment methods than smaller establishments. Thus while large employers used advertising as a recruitment method more frequently than small employers they also used informal methods more frequently. In fact, for large employers informal methods were the most frequently used method of employing both categories of manual workers. However a common pattern appears for small, medium, and large employers in the recruitment of non-manual workers, with advertising being the most frequently used method followed by other formal methods and by informal methods.

A comparison of public and private-sector establishments shows striking differences in recruitment methods in all three skill groups. Public-sector establishments relied much less on informal recruitment methods than did establishments in the private sector.

In addition to recruitment methods, we have examined the extent to which selection and appointments procedures have been formalized through, for example, the use of application forms and through interviewing of applicants prior to appointment. We have collected a good deal of information about such matters with respect to various groups of employees and for the purposes of this analysis we have decided to group establishments into two distinct categories. We have described as formal employers those establishments which subscribe to the following selection and appointments procedures: (i) the employer has a standard application form for vacancies; (ii) the application form is used in the case of all vacancies that arise for that category of employee; (iii) a face-to-face interview is held with the applicant prior to appointment; and (iv) the interview is conducted by more than one person. Establishments which do not meet these four criteria have been referred to as informal employers of that particular category of labour.

On this definition only 32% of workplaces employing skilled manual workers could be described as formal employers of skilled labour. A striking contrast exists between the public and private sectors: public-sector establishments are three times more likely to have formal appointments procedures than private-sector establishments.

Turning to employers of unskilled and semi-skilled manual labour, a similar general pattern exists with only 33% of all establishments meeting the criteria for description as a formal employer and with the same magnitude of difference existing between small and large employers. However, even greater differences between public and private-sector establishments exist, with the public sector being almost five times more likely than the private to be formal employers of this category of labour.

A much higher proportion of establishments (46%) can be described as formal employers of non-manual labour. Nearly all (92%) public-sector establishments as opposed to less than one-third of private-sector establishments meet the formal employer criteria for non-manual labour.

## THE IMPACT OF THE FAIR EMPLOYMENT AGENCY ON EMPLOYMENT PRACTICES

Finally we consider the extent to which the Fair Employment Agency has been successful in bringing about equal oppor-tunity policies and practices. This success of the Agency can be evaluated on many different levels, on not all of which we have gathered information, and therefore no complete answer to this question can be provided here. First, there is the question of how extensively the Agency is used by employers for advice and information. We have already shown that only a very small proportion of employers had initiated voluntary contacts with the FEA and in this respect employers are failing to use the

Agency as a resource centre for advice on fair employment matters.

Secondly, since a major role of the Agency is to conduct strategic investigations of employers' policies and practices under Section 12 of the Fair Employment Act (see Chapters 8 and 9), it would be possible to examine whether such investigations had been successful in bringing about greater equality of opportunity within the organizations investigated. This question is one which can only be comprehensively addressed by a specifically designed study which would ideally include all employers who have been investigated by the FEA. This we have not been able to do. We did however, conduct in-depth case studies of a small group of employers and this group included four who had had formal contact with the Agency involving an enquiry of the Section 12 type.

Our findings suggest that the consequences of formal contact with the Agency vary a great deal from one employer to another and that progress towards equality of opportunity had proceeded at a variable pace for the four employers in question. It was clear that one employer had decided to ignore altogether the Agency's recommendation to carry out a monitoring exercise and the FEA had not, as it has done with other employers, carried out such an exercise using its own staff. A second Agency investigation involved only a limited examination of recent recruitment patterns but no general monitoring of the current work-force had been conducted nor were any recommendations made that such an exercise should be carried out as a matter of course or on a regular basis. Neither of these two employers had had any further contact with the Agency following investigation. In a third organization an effective affirmative action programme had been set up following a compositional analysis of the work-force conducted by FEA staff. Recruitment methods had changed, with advertising of vacancies outside the plant, and in the space of two years the composition of the work-force had altered significantly. In a fourth case little progress had been made in changing the balance of the work-force despite an affirmative action programme agreed with the FEA which involved advertising of vacancies outside the plant and a monitoring of

applications received and appointments made. A major problem for this employer which the affirmative action programme
has not been able to overcome was its loyalist reputation and
geographical location, which made it unattractive to Catholics
as a place to work.

From our case-studies of these four employers it seemed that
formal Agency investigations had a variable and in some cases
limited impact in bringing about greater equality of opportunity. One reason for this is that there appeared to be no
standardized minimum programme of action which the FEA
agreed with all employers following investigation. In some
investigations the FEA carried out an analysis of work-force
composition but in others it did not; some investigations resulted in recommendations to establish monitoring programmes
but others did not; the results of some investigations were
published while the results of others were not. Another reason
for variable success was the failure of employers to implement
agreed programmes of action and the apparent inability of the
FEA to deal with this kind of intransigence.

A third way of assessing the impact of the Agency and
determining its success is to look at the impact Section 12
investigations have had on employers in related industries,
since it might be expected that an adverse report by the Agency
on one employer would cause other employers in similar
situations to reflect on and examine their own procedures. Our
findings show that overall awareness of the Agency is high, but
that only 50% of employers could name any of the organizations investigated by the Agency. However, about 75% of large
employers, that is, those who might be thought of as the target
group of the Agency, could name an Agency investigation.

We sought to examine whether FEA investigations had had
any ripple effect by asking employers whether they had made
any changes to their own policies and practices following an
Agency investigation into another organization. Responses
showed very little evidence of any direct links between FEA
investigations and the actions of other employers. Only 3% of
employers (but 9% of large employers) said that an Agency
investigation had had an impact on policies and practices
within their organization. It therefore seems reasonable to

conclude that, in general, Agency investigations, in so far as they had an impact at all, only had an impact on the employer who is the subject of the investigation and not on any wider community of related employers. In other words employers were not voluntarily taking up or learning from recommendations or suggestions about good practice made by the Agency to employers in similar positions to themselves.

## CONCLUSIONS

The question underlying our research has been one about the contribution an equality of opportunity package can make to undermining job discrimination and thereby reducing structural inequality. Due to the persistence of the sectarian divide, Northern Ireland qualifies as an ideal site for the empirical examination of this question. However, the Northern Ireland experience provides only limited material for answering our question because, with the exception of a few well publicized cases, there has been no general attempt by employers to embrace equality of opportunity as an organizational goal, very few incentives for them to do so, and little by way of sanction against those who have failed to implement affirmative action as a way of eradicating imbalances in the religious composition of work-forces. Northern Ireland's experiment with religious equality of opportunity in employment has, it seems, been rather half-hearted and ineffectual. It is therefore more likely that any improvements in the employment conditions of the Catholic community, or in sections of it, have come about as a result of underlying shifts in patterns of employment across the Province and, in particular, as a result of the expansion of public-sector employment opportunities. This is a question which has been addressed empirically by other researchers (see Chapter 2).

The implementation of an equality of opportunity policy in the midst of civil and constitutional conflict has meant that the policy itself has deeply political connotations. On the one hand, some interpret the recent conversion of the British government to a more robust approach towards equality of opportunity as merely a cynical attempt to win back the allegiance of a large part of the Catholic community to constitutional politics, and at the same time to stave off the threat of American disinvestment in the Province. On the other hand, others, usually from the Unionist camp, see any extension of equality of opportunity as an attempt to undermine the 'natural' in-built majority of the Protestant community by securing or reserving positions for nationalists, many of whom are considered to have no long-term commitment to the Northern Irish state. Thus Unionists are unlikely to embrace whole-heartedly an equality of opportunity policy because they remain deeply suspicious about the longer-term political and constitutional aims of the British government. Because the policy has itself become embroiled in the wider conflict and is seen by both sides as related to larger political strategies, it has not been possible to rely on voluntary compliance or on a hope that the policy's underlying good sense would become apparent as time went by. In consequence, if the British government is serious in its commitment to the implementation of religious equality of opportunity, it is likely that a much greater degree of enforcement and use of sanctions will be necessary.

Despite the fact that the present religious and political conflict has been central to life in Northern Ireland since at least the early 1970s and despite the entrenchment of religious segregation during that period in direct response to political violence and civil disturbance, there is a considerable reluctance to acknowledge that religion is a dynamic force in distributing and allocating economic and social rewards. One reason that equality of opportunity has not taken root has been due to a reluctance to face up to the pervasiveness of the religious divide. Equality policies in Northern Ireland have since the early 1980s been operated on the principle that a person's religion should be of no interest to the employer, and a sea change in attitudes is required if employers are now to appreciate that an explicit awareness of the religion of their employees and job applicants is central to the creation of a fairer and more egalitarian society.

In addition, there has been a failure to set medium-term targets and goals which would serve as measures of success and as indicators of what a policy of equality of opportunity in employment looks like on the ground. Thus there has been an emphasis on the need for statements and declarations of principle but little attention to the practical question of numbers and balance which is the only real indicator of whether a policy has been implemented. The issue is ultimately about a better distribution of jobs for Catholics and about equitable ways of arriving at that aim. The virtual absence from work-places in Northern Ireland of policies and practices designed to create employment equity is then no more than a reflection of the timidity with which the government's *Guide* approached the issue in 1978. It stated: 'The Guide cannot cater for each particular circumstance. Employers and vocational organizations must decide for themselves how equality and opportunity can best be afforded in their particular organization' (DMS, 1978: 5). Given such lack of direction and with no obligation to carry out affirmative action written into legislation, our findings are not altogether unexpected or surprising.

# 7

## Equal Opportunities in the Northern Ireland Civil Service

*Jeremy J. M. Harbison and William J. Hodges*

### INTRODUCTION

The Northern Ireland Civil Service (NICS) is a major employer within the Province. In 1987 it employed over 23,000 non-industrial staff and more than 5,700 industrials; its total work-force of just over 29,000 accounted for around 6% of all employees in employment in Northern Ireland.

As well as being a major employer in its own right, the NICS holds a centrally important position in a divided community where employment opportunities are limited, and where differentials in such opportunities are identified as a key issue dividing the communities (Smith, 1987a). The government's commitment to equality of opportunity in employment needs to be fully reflected in the practices and experience of the NICS, which must be able to demonstrate that it is an equal opportunity employer and is fully representative of the community it serves.

### BACKGROUND

The Fair Employment Agency (FEA) determined in 1980 to investigate the religious structure of employment among non-industrial civil servants to the NICS. The investigation concentrated on the situation in the NICS in 1980, and the Agency published a report on its findings late in 1983 (FEA, 1983). This report showed that whilst for most recruitment grades (that is those grades at which officers normally enter the

The views expressed in this chapter are those of the authors and should not be taken as reflecting the views of the Department of Finance and Personnel.

Service) the NICS had in recent years—i.e. from the mid-1960s onwards—been successfully drawing from the entire labour market, there remained a disparity in relation to the religious composition of the most senior grades in the organization. The FEA report also identified the failure of the Service to attract suitable Catholics with appropriate qualifications for appointments at Executive Officer II level as particularly important, and recommended that more Catholics with good academic qualifications should be encouraged to seek employment within the NICS. The importance of the location of government offices for facilitating equal employment access was noted and the FEA further stressed that the particular religious disparity in relation to the composition of the most senior grades merited special attention. Finally, the FEA recommended that a monitoring system should be established by the NICS to extend the analysis which had been undertaken for this investigation.

The FEA report was welcomed by the government and its recommendations were accepted. The Service subsequently issued an Equal Opportunities Policy Statement in December 1984. This emphasized the commitment of the NICS to equality of opportunity based on the merit principle for all its staff and affirmed that both management and unions would rigorously observe the principles and actively pursue the objectives set out in this Statement. The Statement accepted that words were insufficient to ensure that equality of opportunity in employment was actually operating. The Service needed to be certain that it was operating as an equal opportunity employer and to do so it had to monitor the policy through a detailed information system. An Equal Opportunities Unit was therefore established within the Department of Finance and Personnel with the responsibility for developing, formulating, and co-ordinating equal opportunities policy and practice for the NICS as a whole.

The FEA report had noted that sexual equality was already accepted as a key element of personnel policy with the Service. A Joint Review Group had considered the development of employment opportunities for women in the UK Civil Service in 1982 and this led to an agreed programme of action. Corresponding action was agreed for the NICS in consultation with the Equal Opportunities Commission (NI).

Concurrently with developments relating to sexual and religious employment equality, the Civil Services in both Great Britain and Northern Ireland had extensively revised their codes of practice on the employment of disabled people to ensure that there was no discrimination against them.

The Equal Opportunities Unit (EOU) thus assumed responsibility for ensuring that all persons in the NICS should enjoy full equality of opportunity based on their merit irrespective of their sex, religious affiliation, or any disability. In early 1985 a copy of the Equal Opportunities Policy Statement was issued to every member of staff throughout the NICS with the explicit commitment that 'all eligible persons shall have equal opportunity for employment and advancement in the NICS on the basis of their ability, qualifications and aptitude for the work'.

The EOU proceeded to establish a detailed monitoring system and in July 1986 published its First Report. As the system has developed, further information has been made available through the 1986 Report of the Civil Service Commissioners and the Second Report of the EOU published in December 1987.

## PRINCIPLES OF EQUALITY OF EMPLOYMENT

The policy and practice of the EOU within the NICS has been based on a number of key considerations. The fundamental principle is that all eligible persons shall have equal opportunity for employment and advancement in the NICS on the basis of their ability, qualifications, and aptitude for the work. Such equality in employment means that no one should be denied employment opportunities for reasons that have nothing to do with ability. It entails equal access, free from arbitrary obstructions. Within this context any practices or attitudes that have, whether by design or impact, the effect of limiting an individual's or a group's right to the opportunities generally available because of characteristics attributed to them must be identified and eliminated. It is irrelevant whether such barriers are intentionally or accidentally produced. If any barrier is identified which affects certain groups of individuals in a disproportionately negative way, it is an indication that practices leading to this adverse effect may be inhibiting full equality of employment. What then impedes full access to employment is not the individual's capacity but an external barrier unnecessarily inhibiting such access.

This need to examine the processes of an organization highlights the importance of monitoring. Through monitoring it is possible to identify barriers which are inequitable and may be impeding access or opportunity. Regular monitoring reveals whether, for example, a group has a smaller share of jobs or recruits than one would expect, other things being equal, by reason of its participation in the relevant labour market, and whether a group plays a less important part within the organization than, other things being equal, its share of the workforce within the organization would lead one to expect.

## METHODOLOGICAL ISSUES

Although the equal opportunity developments described in this chapter arose from an investigation centred on the religious background of officers within the Service, the opportunity was taken to integrate the gender and disability dimensions of equality of opportunity into a single and coherent NICS programme. The policy and progamme evolved by the NICS centred on ensuring that no artificial barriers to equality of opportunity based on non-relevant attributes of applicants of employees would be tolerated. In theory, the number and range of non-relevant attributes is considerable, covering such areas as religion, race, age, or sex. With regard to the particular requirements of the Northern Ireland situation, three major areas were initially selected for inclusion in the NICS Equal Opportunity Programme. These were sex and marital status, religious and communal background, and disability. The multi-dimensional approach to the monitoring and evaluation of equal opportunity within the Service, however, lends itself to an extension which could incorporate such other attributes as may be identified as important at a later date.

Regular monitoring enables the organization to examine whether it is attracting a disproportionate share of applicants from a particular group, whether the composition of successful applicants broadly reflects the composition of the original field of applicants, and whether members of any particular group do better than members of other groups in terms of promotion and advancement within the organization. It is the essential first step in helping employers assess whether their recruitment and personnel actually promote equality of opportunity or not.

It is through monitoring that an employer can be informed of the extent to which there are deficiencies in recruitment or promotion processes in the organization which must be remedied, or, alternatively, can learn whether any imbalances are due to circumstances outside its control. The NICS therefore identified the need for the introduction of a comprehensive monitoring system as the foundation upon which its equal opportunities programme should progress. The system should be able to indicate the number of staff employed, how they are distributed throughout a range of different departments, and the type of work carried out, and be able to provide a detailed knowledge of the hierarchical infrastructure—the routes via which employees can enter the system and progress through it, the mechanisms for advancement, and the various means of leaving it.

The NICS is a large multi-disciplinary organization employing staff in a diverse range of posts, for example from senior administrators through specialist scientists to messengers and industrial workers. Two main groups of staff can be identified within the NICS work-force, the larger group of over 23,000 non-industrial staff and a smaller group of 5,700 industrial employees. The non-industrial Civil Service includes administrative, executive, and clerical staff, professional technical and scientific officers, and a range of other miscellaneous grades. The main occupational grouping within the non-industrial Civil Service is the 'General Service' group, comprising in 1987 just under 13,000 staff and embracing the administrative, executive, and clerical grades and supporting staff.

A key characteristic of the non-industrial Civil Service, in all

its occupational groupings, is its hierarchical nature. To facilitate an understanding of the structure and subsequent equal
opportunity analysis, all non-industrial grades can be grouped
into eight salary bandings, first introduced by the FEA in its
1983 Report and developed subsequently by the EOU. All non-
General Service grades can be equated with the grade levels of
the General Service on the basis of salary scales. This produces
eight bandings ranging in descending order of seniority from
Senior Principal and above to Administrative Assistant.

Entry to the system is restricted to relatively few points, the
bulk of recruitment taking place at Administrative Officer and
Administrative Assistant or equivalent levels. However, recruitment on a smaller scale occurs at Executive Officer II level
or above, with the highest general level of Administrative
Trainee reserved for outstanding honours graduates. The
career patterns and promotion expectations are quite different for each grade entry and thus monitoring of the system must
take note of entry level. In particular those entering at
Executive Officer II level and Administrative Trainee tend
to follow an accelerated career path.

In contrast, the industrial staff groupings are considerably
simpler in composition. Whilst industrial staff are involved
in a range of functions, they can be classified into one of
just three industrial grades: supervisory, craft, and non-
craft.

Recruitment to the non-industrial NICS is carried out by a
central agency, the Civil Service Commission, which organizes
a number of competitions to select eligible candidates for
recommendation for appointment. Currently the Commission
receives some 20,000 applications annually in response to
advertised competitions for around 1,500 posts. Comprehensive monitoring must therefore of necessity encompass the
work of the Commission, which operates to standards and
criteria decided by the six Civil Service Commissioners, of
whom two are non-civil servants.

The initial work on establishing an equal opportunities
monitoring framework concentrated on the larger group of
non-industrial staff. For these staff a computerized personnel
information system was already in existence. This system

included, as well as data on age, date and level of entry into the
Service, current level, and area of posting, information on the
sex of each officer. The system also contained information on
whether an officer was registered as a disabled person, or had
been so registered at some time in his or her service career.
Though it was recognized that this information had major
deficiencies for the initial database, it was accepted as an
indicator of the number of disabled within the Service.

No information was held by the NICS on the religion of
individual members of staff. Detailed consideration was given
to the approach to be followed for the collection and maintenance of the religious affiliation data. No monitoring or
evaluation of equal opportunities within the organization
would be possible without almost complete coverage of the
Service, but it was realized that personal religious belief was a
sensitive and important issue for many staff, who might object
to any investigation of their beliefs as an invasion of privacy.
Additionally, since the Service was concerned in organizational
terms with evaluating the overall pattern of equality of opportunity at different levels on the broad Catholic/Protestant
dimension of religious affiliation and communal background, it
was more important to have a statistically reliable method of
assessing these overall patterns than to pursue subjective and
personal aspects of an individual's private beliefs. Finally,
issues such as individual change in religious faith or belief, lack
of religious belief, or antipathy to being classified as a member
of a particular religious grouping were all foreseen as major
problems. Evidence was available from other investigations
that directly seeking information on religious affiliation of staff
could produce unacceptable levels of response.

After detailed consideration and discussion with the Fair
Employment Agency, academics working in the area within
Northern Ireland, and the Service's consultant advisers (the
Institute of Manpower Studies), the decision was taken to use
primary-schooling data as a proxy for religious affiliation. This
decision was based on the rationale that within Northern
Ireland there exist, in the main, two school systems. Some
schools are under the management of the Catholic church and
are attended mainly by Catholic pupils. Other schools are

managed by a variety of bodies including Education and Library Boards and Boards of Governors, and while they are open to pupils of all denominations they are attended mainly by Protestant pupils. A knowledge of the schools attended gives a sufficiently accurate proxy of the religious and communal background of individuals for the purpose of monitoring the overall patterns of employment in the work-force as a whole or at major levels within it.

The extent of cross-over between the two systems has been assessed by a number of researchers, e.g. Murray (1985), and Osborne (1985). All have agreed that the overwhelming majority of pupils attend schools which are only attended by pupils of the same religious groupings. Such information as is available (Murray, 1985; Livingstone, 1987) further demonstrates that, whilst post-primary schools are a good proxy for religious affiliation, the least cross-over between the religious groups occurs at primary-school level.

Having agreed the approach to the collection of information on religious affiliation, the next step was to implement the collection. From the outset the management has accepted that effective equal opportunity policy and practice can only operate with the full support of staff. A joint programme was therefore embarked upon to convince the 23,000 non-industrial staff that the equal opportunity programme must have their support. While the management conducted a series of briefings on the joint policy statement and broad programme negotiated, the trade union side took similar action in relation to their branch chairmen and secretaries. The exercise concluded with a meeting of all branch chairmen and secretaries throughout Northern Ireland being addressed by both trade union and management representatives. The result was a substantial vindication of this line in that almost 80% responded voluntarily to a questionnaire seeking information on schooling. This left only a minority for whom the information had to be extracted from personnel files, resulting in the current situation where the religious and communal affiliation, in terms of schools attended, is known for 95% of the non-industrial service. A similar process for collection of schooling information was introduced at the recruitment stage so that every applicant for a post in the Civil Service co-operates in the identification of religious and communal background.

At an early stage a Code of Practice was drawn up which amongst other things limited access to the information to a small number of staff employed in the EOU, and made a commitment that the religious affiliation data would never be held on a personal file or any other departmental record. The computerized record, within the monitoring system, was the sole record of religious affiliation and any output from this system which included information on religious affiliation would always be in the form of counts, tabulations, or other statistical summaries and of such a nature that individuals could not be identified.

The outcome of all this was the creation of a factual base from which to determine what the actual position was within the NICS as a whole, by department, by occupational group, or by any other subdivision agreed, and on the multi-dimensional basis of sex, religion, and disability.

In such a large and complex organization, the targeting of monitoring and evaluation of equality of opportunity was important. Traditionally, detection of any lack of equality has been based on analysis of labour stocks to a given point in time. However, in organizations with hierarchical seniority systems, the staff composition at any one time will be determined by historical personnel practices the effects of which are currently present and will continue for some time into the future. To monitor equality of opportunity emphasis must therefore be placed not on the composition of the Service ('the stocks'), as has been traditional practice, but on the dynamic nature of the flows into, through, and out of the organization, that is recruitment, promotion, and wastage. The flow rates and thus the potential for compositional change are determined by factors such as wastage and whether the organization is in a state of expansion or contraction. Additionally, the possibility of entering the Service at different levels and into career streams with differential rates of movement emphasizes the dynamic nature of flows and the complexity of their potential for organizational change. Compositional analysis of an organization thus constitutes a sensitive barometer to change and establishes that monitoring must target, in particular, recruitment into the organization and promotion within it.

The strategy developed by the EOU was, therefore, an initial examination of the current composition of the non-industrial Service (which could be done rapidly upon completion of the data collection exercise). Concurrently, the EOU established systems to ensure that flow information relating to recruitment, promotion, and wastage to, within, and from the NICS would be collected. The next stage in this strategy involved analysis of these dynamic components of change. Finally, the EOU set in operation a similar pattern to ensure coverage of the industrial component of the NICS.

It was further appreciated, from an early stage, that availability of raw material on equality of employment opportunity data relating to sex, religious affiliation, and disability was a necessary but not sufficient base to establish whether policies and practice of equal opportunity were operating in practice. Detailed work carried out during the Fair Employment Agency investigation (FEA, 1983) highlighted the importance of the range of information which had to be considered in any valid assessment of the organization's practices and the complexity of analysis required to comprehend its dynamics. For example, analysis has demonstrated that individuals differ not only in sex or religious terms but in possession of qualifications, in level of entry to the Service, or in occupational group entered, and that a complex pattern exists between the equal opportunity variables and these work-related characteristics which influences progression through the Service. Professional statistical staff within the NICS have been developing approaches to the handling and analysis of the data, as well as establishing the software and hardware systems required. This work is detailed in a paper by Stevenson *et al.* (1988).

## CURRENT POSITION

The recent position regarding equal opportunities in the Service has been documented by the publication of the Second Report of the EOU and the 1986 Civil Service Commissioners' Report. Compositional information is now fully available for the total service of over 29,000 industrial and non-industrial employees. For the non-industrial group, full information is held by department, by occupational group, by level of Officer, and by age, length of service, qualifications held, and a number of other characteristics. All this information is available dis-aggregated in terms of sex, religious affiliation, and disability. Rather less detailed information is available for the industrial staff, but data can be provided in terms of department, grade, age structure, and functional analysis for the equal opportunity dimensions of sex, religious affiliation, and disability. Tables 7.1 and 7.2 show the type of data that is available to record equal opportunities in terms of sex and religious affiliation. The figures suggest that whilst the composition of the non-industrial service appears broadly representative of the Northern Ireland community in overall terms, women and

TABLE 7.1. *Non-industrial Civil Service: by grade level and gender, January 1987* (Row %)

| Level[a] | Sex | | | | Total |
|---|---|---|---|---|---|
| | Male | | Female | | |
| | N | % | N | % | |
| A | 499 | 90.9 | 50 | 9.1 | 549 |
| B | 681 | 93.5 | 47 | 6.5 | 728 |
| C | 1,401 | 89.5 | 163 | 10.4 | 1,564 |
| D | 1,612 | 81.2 | 372 | 18.7 | 1,984 |
| E | 2,428 | 74.6 | 824 | 25.3 | 3,252 |
| F | 2,342 | 56.8 | 1,780 | 43.2 | 4,122 |
| G | 1,686 | 25.2 | 5,030 | 74.8 | 6,716 |
| H | 1,096 | 25.2 | 3,259 | 74.7 | 4,355 |
| TOTAL | 11,745 | 50.5 | 11,525 | 49.5 | 23,270 |

[a] All non-industrial grades have been grouped into 8 salary bandings, equated with the grade levels of the general service on the basis of salary scales. The bandings are A: Senior Principal and above; B: Principal; C: Deputy Principal; D: Staff Officer; E: Executive Officer I; F: Executive Officer II; G: Administrative Officer; H: Administrative Assistant.

*Source*: Equal Opportunities Unit, 1987.

Catholics remain under-represented at more senior levels of the organization. As noted above, however, such information is of relatively limited value in any comprehensive evaluation of equality of opportunity in current terms.

Information on the dynamic components of the service is now starting to emerge. At recruitment level applications from almost 30,000 individuals for non-industrial posts have been analysed in equal opportunity terms and by broad occupational grouping for the period August 1985 to December 1986. Detailed information on applications and appointments for 118 completed recruitment competitions is available and is shown in summarized form in Tables 7.3 and 7.4.

A number of points can be made concerning this information. One is the large number of applications for available appointments—a ratio of more than 13 to 1, indicative of the job position in the Province on the basis of currently available information. The Civil Service Commission would appear to be connecting satisfactorily with the total labour force in Northern Ireland when the characteristics of applicants and appointments are considered. The value of the use of schooling information in determining religious affiliation is clear, with only 5% of applicants and under 3% of appointments being classified either as educated outside Northern Ireland or as individuals for whom educational information is not known. What is also clear is the variation in the composition of applications in certain categories in equal opportunity terms. For example 61% of applicants for general service grades were female in contrast to 7% for posts in the technology grades. Similarly almost 49% of applicants for general service posts were Catholic in contrast to 32% for posts in the scientific grades. Appointments in the main mirror applications, females having a success rate of 8.6% compared to 6.4% for males, Protestants having an overall success rate of 8.1% compared with 7.0% for Catholics. In this period only 237 applications were received from persons registered as disabled, and 4 appointments were made.

The variability in applications from different sections of the labour market is interesting. Until accurate and comprehensive labour availability estimates of the number of

TABLE 7.2. *Non-industrial Civil Service: by grade level and religion, January 1987* (Row %)

| Level[a] | Religion | | | | | | | | Total |
|---|---|---|---|---|---|---|---|---|---|
| | Protestant | | Catholic | | XNI[b] | | Not known | | |
| | N | % | N | % | N | % | N | % | |
| A | 384 | 69.9 | 79 | 14.4 | 67 | 12.2 | 19 | 3.5 | 549 |
| B | 513 | 70.5 | 108 | 14.8 | 90 | 12.4 | 17 | 2.3 | 728 |
| C | 1,077 | 68.9 | 300 | 19.1 | 111 | 7.1 | 76 | 4.9 | 1,564 |
| D | 1,272 | 64.1 | 552 | 27.8 | 54 | 2.7 | 106 | 5.3 | 1,984 |
| E | 1,910 | 58.7 | 960 | 29.5 | 91 | 2.8 | 291 | 8.9 | 3,252 |
| F | 2,327 | 56.4 | 1,395 | 33.8 | 90 | 2.8 | 310 | 7.5 | 4,122 |
| G | 3,557 | 52.9 | 2,913 | 43.4 | 120 | 1.8 | 126 | 1.9 | 6,716 |
| H | 2,486 | 57.0 | 1,542 | 35.4 | 94 | 2.2 | 233 | 5.4 | 4,355 |
| TOTAL | 13,526 | 58.1 | 7,849 | 33.7 | 717 | 3.1 | 1,178 | 5.1 | 23,270 |

[a] All non-industrial grades have been grouped into 8 salary bandings, equated with the grade levels of the general service on the basis of salary scales. The bandings are A: Senior Principal and above; B: Principal; C: Deputy Principal; D: Staff Officer; E: Executive Officer I; F: Executive Officer II; G: Administrative Officer; H: Administrative Assistant.

[b] XNI means schooling outside Northern Ireland.

*Source:* Equal Opportunities Unit, 1987.

TABLE 7.3. *Applications for completed competitions by recruitment category, sex, and religion (August 1985–December 1986)*

| | Number of competitions | Male | Female | Prot- estant | Cath- olic | XNI and not known | Total |
|---|---|---|---|---|---|---|---|
| General service grades | 5 | 4,684 | 7,302 | 5,853 | 5,828 | 305 | 11,986 |
| Scientific grades | 15 | 1,015 | 461 | 843 | 470 | 163 | 1,476 |
| Technology group | 20 | 773 | 62 | 451 | 346 | 38 | 835 |
| Legal grades | 6 | 210 | 218 | 246 | 163 | 19 | 428 |
| Computer grades | 1 | 81 | 43 | 52 | 63 | 9 | 124 |
| Other professional and departmental grades | 52 | 2,074 | 485 | 1,597 | 651 | 311 | 2,559 |
| Centralized services | 19 | 2,099 | 477 | 1,613 | 785 | 178 | 2,576 |
| TOTAL | 118 | 10,926 | 9,048 | 10,655 | 8,306 | 1,023 | 19,984 |
| % | | 54.7 | 45.3 | 53.3 | 41.6 | 5.1 | 100.0 |

*Note*: XNI means schooling outside Northern Ireland.

*Source*: Equal Opportunities Unit, 1987.

TABLE 7.4. *Appointments from completed competitions by recruitment category, sex, and religion (August 1985–December 1986)*

|  | Number of competitions | Male | Female | Protestant | Catholic | XNI and not known | Total |
|---|---|---|---|---|---|---|---|
| General service grades | 5 | 453 | 683 | 649 | 470 | 17 | 1,136 |
| Scientific grades | 15 | 31 | 15 | 27 | 14 | 5 | 46 |
| Technology group | 20 | 82 | 6 | 57 | 29 | 2 | 88 |
| Legal grades | 6 | 13 | 16 | 19 | 10 | 0 | 29 |
| Computer grades | 1 | 15 | 19 | 15 | 15 | 4 | 34 |
| Other professional and departmental grades | 52 | 97 | 33 | 80 | 37 | 13 | 130 |
| Centralized services | 19 | 13 | 7 | 15 | 4 | 1 | 20 |
| TOTAL | 118 | 704 | 779 | 862 | 579 | 42 | 1,483 |
| % |  | 47.5 | 52.5 | 58.1 | 39.0 | 2.8 | 100.0 |

*Note:* XNI means schooling outside Northern Ireland.

*Source:* Equal Opportunities Unit, 1987.

appropriately qualified individuals of working age who are female, Catholic, and disabled can be analysed, it is not possible to establish at these levels of detail whether the NICS is currently connecting fully with all parts of the labour market. On the basis of available evidence, however, the situation appears generally satisfactory for members of groups other than the registered disabled, a situation not unique to the NICS (National Audit Office, 1987).

Individual competitions have also been examined in great detail. One is the Executive Officer II competition, which is of particular importance in that it represents recruitment against more stringent qualification requirements with the expectation that successful candidates will progress to the higher levels of the service. It will be remembered that one of the FEA recommendations related to the failure of the NICS to attract suitably qualified Catholics for appointments at Executive Officer II level.

The results of the various stages in the 1985 Executive Officer II competition are summarized in Table 7.5. What is immediately clear from the table is that whilst the number of Catholics and women who accepted appointment at Executive Officer II level appears to relate to the number of applicants, in-depth examination of the processes involved in recruitment has identified that certain groups of candidates have not performed as well as others on the standardized selection tests. These tests were devised and developed by the United Kingdom Civil Service Commission for all Executive Officer posts in the UK. The three tests comprised an intelligence test, a test requiring candidates to understand and deal with graphical tabulated data, and a short-answer test concerning problems of the kind that arise in administration. A detailed investigation of the competition showed that Catholic females fared particularly poorly on these tests. Of the 304 Catholic females who took the tests only 7 (2.3%) achieved a sufficiently high mark to be included in the group called for interview, in comparison with 6.1% of Catholic males, 12.7% of Protestant females, and 13.1% of Protestant males. Further investigation showed that neither academic attainments (measured in terms of O level grades) nor age could account for the differences

TABLE 7.5. *Summary of the key stages in the Executive Officer II competition (1985), Protestant and Catholic applicants*

| | Total | Protestant | | Catholic | | Prot-estant (%) | Cath-olic (%) | Male (%) | Female (%) |
|---|---|---|---|---|---|---|---|---|---|
| | | Male | Female | Male | Female | | | | |
| Applicants | 1,887 | 476 | 499 | 502 | 410 | 51.7 | 48.3 | 51.8 | 48.2 |
| Qualified applicants | 1,781 | 453 | 474 | 468 | 386 | 52.0 | 48.0 | 51.7 | 48.3 |
| % qualified applicants | 94.4 | 95.2 | 95.0 | 93.2 | 94.1 | | | | |
| Attended test | 1,455 | 388 | 386 | 377 | 304 | 53.2 | 46.8 | 52.6 | 47.4 |
| Eligible for interview on basis of test score | 130 | 51 | 49 | 23 | 7 | 76.9 | 23.1 | 56.9 | 43.1 |
| % eligible for interview | 8.9 | 13.1 | 12.7 | 6.1 | 2.3 | | | | |
| Interviewed | 117 | 45 | 42 | 23 | 7 | 74.4 | 25.6 | 58.1 | 41.9 |
| Passed interview | 34 | 13 | 10 | 9 | 2 | 67.6 | 32.4 | 64.7 | 35.3 |
| % of interviewed who passed | 29.1 | 28.9 | 23.8 | 39.1 | 28.6 | | | | |
| Accepted appointment | 25 | 7 | 7 | 9 | 2 | 56.0 | 44.0 | 64.0 | 36.0 |
| Appointees as % of qualified applicants | 1.4 | 1.5 | 1.5 | 1.9 | 0.5 | | | | |

*Source:* Equal Opportunities Unit, 1987.

between the objective test performance of Catholic (and particularly female Catholic) applicants and Protestant applicants. Other competitions have produced similar results and additional research is under way to explore why this key group appears to perform poorly on carefully developed selection measures.

Monitoring of all promotion boards in the non-industrial service is in progress, but as a board life can be up to 24 months little completed information is presently available. Already, however, a feature which is becoming increasingly evident from monitoring of completed boards is the extent to which some individuals, in particular females, choose not to be considered for promotion. This apparent voluntary withdrawal from consideration has an obvious detrimental effect on the movement of particular groups of staff into higher grades, and detailed research aimed at identifying the factors which influence individuals to withdraw from the promotion process is under way.

Monitoring of wastage from the non-industrial service has demonstrated the different patterns of leaving shown by males and females, and to a lesser extent differences between Catholic and Protestant officers. Other characteristics of staff interact with the equal opportunity dimensions, the more important already identified being age, level of post of officer, and type of occupational group.

### KEY COMPONENTS IN INTRODUCING EQUAL OPPORTUNITIES

Now that the equal opportunity programme has been operational for several years, some comments of a more general nature may be of relevance. Crucial to the introduction and operation of such a programme is full commitment from senior management and unions. The importance of this has already been noted during discussion of the establishment of an equal opportunities database, which moved ahead successfully mainly because of this joint commitment. Circumstances have progressed and, in this dialogue between management and trade unions, there is now available information on recruitment and promotion arrangements and how they are operating, on wastage, on the distribution of staff across the Province, and on much more which was never there before.

This had enabled a joint look to be taken at the effects of agreed policies in all areas of the Civil Service ranging from level of entry and qualifications at recruitment through to seniority requirements for promotion, departmental differences, and what additions need to be made to conditions of service within the NICS to enhance the equality of opportunity programme on which the Service has embarked. An important dimension to this work has been the involvement of the Institute of Manpower Studies so that both staff and management can be completely satisfied that all of the facts recorded were correct and that the statistical conclusions drawn were valid.

The management has found it necessary to update substantially facilities connected with personnel records and to enhance and to update, more rapidly than would otherwise have been the case, the use of information technology. The trade unions have been afforded access on an equal basis with management to the products of these information systems. Since the programme has been embarked upon there has been the introduction by agreement of considerable improvements on the use of flexitime; on part-time working arrangements; on training, where single-sex training courses are being piloted; on examinations of child-care facilities; on the possibilities of career breaks and improved reinstatement opportunities, particularly for those who have to leave on domestic grounds; and new insights on differential wastage rates from differing groups within the Civil Service as a whole.

The introduction of monitoring to the industrial side has again taken place with the whole-hearted co-operation of the major industrial trade unions. For industrial staff a parallel course to that successfully deployed in respect of non-industrial staff has been initiated. Meetings were held with groups of industrial staff on their work sites all over Northern Ireland which were attended jointly by management and trade union representatives. This process has led to the attainment of an even higher response rate on information collection than was achieved for non-industrial civil servants.

Associated with this openness in communication and the co-operative approach to the introduction and implementation of equal opportunities has been the production of a Code of Practice with absolute guarantees that the religious affiliation data were confidential and would never be used on an individual basis or be available on individual personal files.

An additional requirement which is of considerable importance is the commitment to publish the results of the equal opportunities monitoring, 'warts and all'. The systems that have been established will probably over time enable a closer and more intimate examination of NICS practices to be made than is the case for any other major organization, certainly in the UK. Already the examination has identified a number of potential problems: the position at recruitment level identified in the key Executive Officer competition has been noted. However, for the NICS to be able to demonstrate that only qualified people can be appointed to the NICS and equally that it is striving to identify and eliminate barriers to such employment affecting members of any group, openness and completeness in reporting to the community it serves is essential.

The operation of an effective equal opportunities programme also requires the ability to take action where barriers are identified or unacceptable practice uncovered. This is another area where full senior management and trade union commitment is essential. The EOU has already, through a series of actions ranging from commissioning additional research on potential problem areas, to working with departments, training officers and unions on altered current practice, and issuing formal advice by means of Civil Service Circulars, altered personnel policies and practice within the Service in a manner designed to facilitate equality of opportunity.

A final comment to make is that for the successful introduction of an equal opportunities programme in a large organization, adequate resources are required. Not only are comprehensive personnel systems needed within the organization upon which to develop such a programme, but within the NICS additional resources were involved including the staffing of the EOU, professional resources from the Policy Planning and Research Unit to work with the EOU in the implementation of monitoring and analysis of information, hardware and software resources, and, not least, the time and availability of officers in the Civil Service Commission, in personnel divisions of all departments and in many other areas to incorporate equal opportunities into the human resource planning and management process. The decision to allocate such resources to equal opportunities is obviously a major one for any organization, but it is a necessary step for an effective equal opportunities policy.

## CONCLUSIONS

By 1988 the Northern Ireland Civil Service had introduced a comprehensive equal opportunities programme for both industrial and non-industrial staff. Recruitment is being monitored, as is promotion and wastage. Special research exercises and investigations are under way in a number of areas. There remains a major work programme to refine and extend the work. The DED Consultative Paper (1986) identified the Civil Service equal employment model as a suitable model to be followed by the wider public sector, and the details of the new Fair Employment (Northern Ireland) Act have made it clear that in future all employers in the public sector (as well as the private) will be required to actively practice equality of opportunity in employment. One component of this will require all employers with more than ten employees to carry out annual monitoring of their work-force, which will cover over 200,000 jobs in the Province.

The work to date has highlighted the importance of both the practical and the philosophical approach to equal opportunities; if equal opportunities are to be incorporated as part of human resource management in individual organizations this can best be done through a multi-dimensional approach. Evidence accruing from the current work strongly reinforces this view—for example issues highlighted earlier at recruitment are not related solely to gender or to religious affiliation, but are very much characteristic of a specific group defined in terms of both sex and religious affiliation.

Some recent data has become available about the community's perception of the NICS. A study by Smith (1987*b*) records the perceptions and views of a large sample of the population of Northern Ireland on various issues of equality and inequality. The sample (of round 1,700 individuals) was asked its views on discrimination in employment. Respondents recorded their views on four kinds of employers: small businesses, large businesses, local councils, and the Civil Service. Views on the prevalance of discrimination in recruitment to employment varied widely depending on the kind of employer considered. A majority of respondents (63%) thought there were small businesses that discriminate, substantial proportions thought there were larger businesses (40%) and local councils (39%) that discriminate, but the proportion who thought the Civil Service discriminated was comparatively small (17%). Whilst in the particular context of Northern Ireland it is reassuring that the public appear to perceive the NICS as, generally, a fair employer, the size of the group which still believes discrimination occurs indicates the need for the NICS to continue to seek to ensure and to demonstrate that its practices are fair to all.

# 8

## The Role of the Fair
## Employment Agency

*Robert G. Cooper*


### INTRODUCTION

When the Fair Employment Agency (FEA) was set up in 1976 Catholics in Northern Ireland were grossly under-represented in the public sector and, in particular, almost non-existent in senior parts of that sector. They were over-represented in the construction industry and in some of the services, particularly those dealing with the leisure industry. They were under-represented in manufacturing industry. Overall, they tended to be well represented in lower-paid occupations, but poorly represented in higher-paid occupations. They were under-represented in the better-paid parts of the service industries, such as banks, building societies, and insurance companies. They were two and a half times as likely to be unemployed as Protestants (see Chapter 2).


### EDUCATION AND PROMOTION

When the Agency was set up it concentrated, for the first year of its existence, on education and promotion on the one hand, and on the investigation of individual complaints on the other. The major problem which the Agency faced in the education field was that, at that particular time, the general consensus of opinion among employers had been strongly shaped into thinking that employers should know as little as possible about the religion of their employees. The stereotyped view of the Northern Ireland man was that he could never relax until he had found out the religion of the person he was talking to, and that he would therefore go to endless trouble to try to find it out. As Seamus Heaney (1990) described it:

> Smoke-signals are loud-mouthed compared with us:
> Manœuvrings to find out name and school,
> Subtle discrimination by addresses
> With hardly an exception to the rule
>
> Than Norman, Ken and Sidney signalled Prod
> And Seamus (call me Sean) was sure-fire Pape.
> O land of password, handgrip, wink and nod,
> Of open minds as open as a trap.

It had been the normal practice of many employers in the past to ask the religion of potential employees on the application form. The official explanation was that if someone dropped a hammer on the head of an employee, the employer needed to know whether to get a priest or a Protestant clergyman. In not all cases was that false, but on at least some occasions there was room to suspect that the real reason was so that people would know on which heads hammers should be dropped! More recently, employers considered that seeking this information might not be a good idea after all, and believed that the less known about the religion of employees the better.

Application forms were therefore constructed which sought to minimize the amount of information about religion which could be obtained from them. It became quite normal to ask for the type of school but not the name of the school, as the school was recognized as being the most reliable indicator of religion. The public sector took the lead in this practice. On one occasion a Health Board had on the first page of an application form, in enormous block letters, 'TYPE OF SCHOOL, NOT NAME OF SCHOOL', and on the third page asked for two references, one of whom had to be the principal of the school where the candidate was educated. The difficulty about this approach was that only a blank application form which sought no information could successfully conceal the religion of the applicants on many occasions. For example, possible indicators of religion on the application form are Christian name, surname, address, nationality, subjects taken at school, referees, sports, interests, and former employers. From the point of view of the FEA such an approach made the monitoring of equality of opportunity in employment very difficult. If the problem of equality of opportunity was simply that of direct discrimination then it is possible that one could cope without getting involved in detailed monitoring. However, direct discrimination is only a small part of the problem.

## INDIRECT FACTORS

A major problem was and still is the 'chill factor'. It is quite common in Northern Ireland for employees to be reluctant to work in particular companies either because they are afraid that they will be in physical danger or because they fear that they will be discriminated against; in other words the company is identified as a Protestant or Catholic company. The chill factor may have arisen because of past discrimination or because of geographical factors. It may have arisen because of or despite the attitude of the employer. For example, there are two firms in Northern Ireland where there is a strong chill factor acting against Protestants and yet both companies have been owned by Unionist councillors. The report by the Policy Studies Institute (Smith, 1987*a*) showed that a large number of organizations are identified in this way as Protestant or Catholic. It has always been the view of the FEA that employers have a duty to attempt to overcome the chill factor. However, it can only be identified if the labour force is monitored.

Unconscious discrimination is equally a problem. In a typically divided society selectors are more likely to have many points in common with applicants from their own religious group; for example, school, home town, sports played, organization membership. In a situation where there are many suitable candidates for a job, these types of factors can make all the difference in deciding who is selected. The selector will often not be conscious that religion has entered into it, but religion has in fact played a significant part.

A further problem is employment practices and conditions which are not intentionally discriminatory, but which have an adverse impact on one section of the population. For example, word of mouth recruitment with the use of unsolicited applications is probably the single largest factor in the maintenance of inequality, and yet it is still widely used.

## MONITORING

It is not possible to begin to deal with these types of inequality unless there is a proper monitoring system. The first job, therefore, which the FEA had to undertake was to encourage employers to introduce such a system. It would be impossible to over-estimate the shock which this represented to many people. On one occasion a senior executive of a firm met an official of the FEA and apologized for not attending an Agency seminar but said that he had sent the company secretary. However, he said the company secretary was so stupid that he always got hold of the wrong end of the stick and, unbelievably, returned and told the Board that the Agency recommended that employers should ascertain the religion of their employees. The opposition to monitoring ran right across the board, from bigots who felt that it would show up a lot of nasty situations which they would rather keep concealed, through to liberals who genuinely believed that people should not be categorized as Protestant or Catholic; from atheists and agnostics who objected to being described as Protestant or Catholic, to Catholics who had been fighting for so long against discrimination with the argument that religion blindness was the answer that they found it difficult to accept that it was not.

The FEA ran many seminars to convince employers of the necessity of monitoring, and although some employers were convinced intellectually, few, if any, actually did anything about it. The Agency's work in this field was not helped by the fact that the *Guide to Manpower Policy and Practice* (DMS, 1978) issued by the government under the Act, while it recommended monitoring, did so in such a tentative and half-hearted fashion that it did not carry very much conviction. To make matters worse the government, which had issued the *Guide*, recommending, however cautiously, that employers should monitor their labour forces, refused point-blank to follow the same course for its own employees. Indeed, it only embarked on the current monitoring programme after the FEA conducted a formal investigation of the Northern Ireland Civil Service, an investigation which took up much of the Agency's resources for a period of approximately three years (FEA, 1983).

## INVESTIGATIONS

As a result of the response to monitoring, the FEA made a decision that the only way in which employers would be persuaded to monitor their labour force was for the Agency to embark on a substantial programme of investigations. Only when an employer had been 'found guilty' by the Agency of not providing equality of opportunity could the Agency insist that the employer should monitor his labour force. The FEA, as a short cut, decided that it would try to monitor, for periods of six months, the recruitment of all those employers who appeared to be engaged in active recruitment. Four organizations were selected to begin this exercise. Immediately the Agency ran into difficulties, as it was found that, public appearance to the contrary, the volume of recruitment in each of these companies, while large by the standards of certain other employers, was still over a six-month period not sufficient for the Agency to arrive at any major statistical conclusions. The Agency therefore decided to abandon this approach. Of the four companies examined, two were subsequently subject to a full compositional analysis, but the other two, where the general picture seemed to be quite reasonable, were not investigated. These are the only two companies investigated by the Agency where compositional analyses of the employees were not required.

The FEA decided that in future it could not rely on such a short cut. Instead, it would have to embark on full investigations of organizations which would include carrying out compositional analyses. During the life of the FEA, the major areas investigated in this way were the Northern Ireland Civil Service, Northern Ireland Electricity, Northern Ireland Railways, Ulsterbus/Citybus, Northern Ireland Airports, Northern Ireland Housing Executive, two Health Boards and one Education Board, major engineering companies in Belfast, manufacturing companies in Derry, the major banks, the major building societies and eleven insurance companies, the two universities, and seventeen motor agents. The Fair Employment Commission (FEC), as the FEA has become, is currently completing investigations of all twenty-six District Councils.

Just over one-fifth of all employees in Northern Ireland are in companies which have been, or are being, investigated by the FEA/FEC. One other important aspect of the investigation work which needs to be emphasized is that unless the FEA made a formal finding that a company was not providing equality of opportunity in employment it had no power to insist that the company initiate a monitoring scheme. It could, and did, recommend that all companies should do so, but only where it had made a finding could it insist on monitoring. On occasion the Agency was criticized for the vagueness of its findings, but the Agency was satisfied if the finding was sufficient to enable it to negotiate a monitoring scheme and, where necessary, an affirmative action agreement.

## THE PUBLIC SECTOR

The changes which have taken place in the public sector in the last decade are fairly substantial; Catholics are now represented in the Civil Service at exactly their proportion of the population as a whole. Since the Agency's investigation the change in the overall composition of the Civil Service has meant that there are 1,700 more Catholics and 400 less Protestants. In the Staff Officer grade the proportion of Catholics had increased from 19% in 1980 to 28% in 1986. Because of the very substantially increased recruitment of Catholics in the late 1970s and 1980s there is still a major discrepancy in the age structure, with more than half the Catholics being under the age of 30 compared to approximately one-third of Protestants. Catholics are still significantly under-represented at higher levels and it will take time to see if this is simply because of the age structure or if there are other problems (see Chapter 7).

The worst area in the public sector in the past was in local government, where the tradition of 'employing one's own' was powerful. This was the area which was identified by the Cameron Commission (1969), which investigated the causes of the early disturbances in Northern Ireland, as exhibiting substantial evidence of discrimination. Initial examination the Agency has undertaken into local councils suggests that the tradition of 'employing one's own' is not wholly dead, although the chill factor rather than deliberate discrimination may now be a larger reason for under-representation. Major changes have, however, taken place.

The FEC has obtained details of the religious composition of employees of all twenty-six District Councils. The proportion of those whose religion has been identified as Catholic employed in the twenty-six councils is 35%, which is about the same as the proportion of Catholics in the population in the areas covered by the councils. There are, however, wide discrepancies in individual councils. Twelve of the councils have employment figures for Catholics or Protestants within 5% of the local adult population. Ten have figures where the variation is more than 5% in favour of Protestants and in four councils it is more than 5% in favour of Catholics. In three Unionist-controlled councils the proportion of Catholics is less than half the proportion in the local area, while in one Nationalist council the proportion of Protestant employees is less than half the proportion of the population. While the overall proportion of Catholics is satisfactory, they are still significantly under-represented at senior level where only 23% of senior employees are Catholic. Usually the number of senior employees is so small that one cannot attach significance to the small number of Catholics in senior positions in any one council but, out of the sixteen councils which since reorganization in 1973 have always been under Unionist control, eleven do not have a single Catholic in a senior position. The Local Government Staff Commission code operated by all councils only permits councillors a decision-making role in the appointment of senior staff.

Most powers, however, were taken away from local government in the early 1970s and given to public boards (Birrell and Murie, 1980). In the public boards which the FEA examined the level of Catholics has improved significantly, and it is quite startling at senior level. Of the nine public boards dealing with Health and Social Services and the Education and Library Service, none in 1972 had Catholic Chief Executives, but five now do.

In 1973 the Housing Executive was given responsibility for all public-sector housing. In 1984, when the FEA investigated the Executive, it found that of the older employees who had been employed before reorganization less than 25% were Roman Catholic but of the total employees 37% were Roman Catholic. By 1988 this had risen to 43% and since 1984 more than 50% of those recruited have been Roman Catholic. Although there has been a decline of some few hundreds in the total employees of the Housing Executive during this period the number of Catholics in employment has increased by 80. The number of Protestants, however, has declined by almost 400.

In other areas in the public sector, such as the Fire Authority and the Ambulance Service, the recruitment of Catholics has now reached a satisfactory level, although it will take some time to work its way through to a substantial level of overall representation of Catholics. In Ulsterbus/Citybus Catholics are somewhat over-represented, while in the Post Office and British Telecom the proportion of Catholics has traditionally been higher than their proportion of the population. In the Electricity Service although improvements have taken place the level of Catholics is still too low, as is also the case in Northern Ireland Railways.

There are still major pockets of the public sector which have not been investigated by the FEA/FEC and for which no figures are yet available (although this will change with the monitoring returns required under the new legislation). The indications would suggest, however, that while Catholics may still be under-represented in the more senior posts of the public sector, particularly the Civil Service and local councils, their overall level of representation cannot be too much out of line with their proportion of the adult population.


## THE PRIVATE SECTOR

With regard to the private service sector, the recruitment of Catholics in the banks, building societies, and insurance companies is approaching a satisfactory level although, since the Agency's investigations, the level of recruitment overall has been quite low and so the rate of change in the overall composition will be slow. When the Agency investigated the building societies in 1986 eight of these had together 14% Catholic employees. The same eight since then have recruited 34%

Catholics. The five banks in 1986 had 29% Catholic employees but have recruited 39% Catholics since then. The major area where too little change is taking place is in the manufacturing industry. We can divide the manufacturing industry into three categories.

These are, firstly, long-established large companies indigenous to Northern Ireland in traditional industries, where discriminatory practices were endemic in the past. Some change has taken place in a number of these companies. For example Shorts, by far the largest manufacturing company in Northern Ireland, has increased its proportion of Catholics from 3–5% when the Agency investigated it, to 11–12% now. It is now the second largest employer of Catholics in manufacturing industry. There has, however, been a massive rundown in employment in most of these companies over the last decade or so. Between 1973 and 1986 employment in those firms declined by over 40% or 27,600 jobs (NIERC, 1989). This has been a major drawback in effecting change.

Secondly, there are manufacturing factories set up by outside investors in the last few decades. The major problems of inequality are not generally connected with these companies, although some of them were content to allow their composition to be shaped by the employment patterns created by earlier indigenous companies rather than to work actively to overcome these patterns. This is an area where one would have hoped that new job opportunities would have become available for Catholics. Unfortunately, this sector declined from 92,000 jobs in 1973 to 42,000 in 1986 so that it was able to make little overall contribution to improving the position of Catholics.

Thirdly, there are very small manufacturing companies where there is a tendency to employ people on the basis of relationship to the working proprietor, senior staff, or other employees; in this sector of employment there is a widespread pattern of 'employing one's own' and since Catholics at present own fewer of these companies proportionately than Protestants, they suffer most from such practices.

Broadly speaking, few companies have been prepared in the past to initiate proper equality of opportunity programmes unless the FEA investigated them, and with the resources which the Agency had in the past it was not possible to devote significant resources to investigating smaller companies.

The overall pattern, therefore, is of significant improvement in the public sector, some improvement in some areas of the private sector, and too little improvement in other areas of private employment, with the major problem being long-established indigenous Northern Ireland industries and smaller companies.

## OVERALL PROFILES

The position of Catholics who are in employment has changed significantly (see Chapter 2). For example, in 1971 it would have needed Catholic managers in large establishments to have increased by 300% to equal the Protestant rate; by 1981 the deficit was 80%. In 1971 an increase of 36.5% would have been needed in the number of Catholic foremen to equal the Protestant rate, and by 1981 this had fallen to 26.4% (Osborne and Cormack, 1987).

Catholics have made very substantial progress in some areas where the level of representation was considerable in the past, such as the legal profession, and where they now have a representation which is somewhat above their proportion of the population. Catholics have increased in such areas as personnel and industrial relations managers, economists and statisticians, professionals in science, engineering, and technology, and as scientists and engineers. Yet, in spite of these improvements, the discrepancy between Catholic and Protestant rates of unemployment remains constant, and that inevitably overshadows all else. To the young Catholic unemployed in West Belfast or Derry, the fact that Catholics in the Civil Service or the public boards, or in jobs such as scientists, statisticians, and professional engineers, may have increased is irrelevant. The harsh reality that he faces is that during the 1980s it has become much more difficult for him to find a job and he suspects that his Protestant counterpart does not have the same difficulty. The evidence suggests that he is right, and his sisters would agree.

## RESULTS

In all the areas the FEA investigated, and where it obtained a monitoring agreement with the employer to report on a regular basis what was happening inside the plant, the Agency found that, with few exceptions, improvements in the position of Catholics took place. Sometimes the improvements were substantial, as in the Ambulance Service, the building societies, and the Fire Service. In other areas they were and remain smaller but still significant. The exceptions remain those areas, particularly in manufacturing firms, where little recruitment has taken place since the FEA's investigation; for example, one manufacturing company investigated by the Agency has had a reduction, during the period since the investigation, from 1,500 employees to 750 employees. As a result, we have the paradoxical position of a significant improvement in the position of those Catholics in employment, and yet the discrepancy between Catholic unemployment and Protestant unemployment has remained quite constant.

The changes in patterns of employment have not been helpful for Catholics. For example, the manufacturing industry, which was a core Protestant industry, has almost halved, but Protestants have been able, to some extent, to cushion this rundown by increases in security industries, in which Catholics find it difficult to seek employment for reasons connected with the present level of violence. In addition, Catholics have suffered badly because the construction industry has collapsed as a major source of employment. Hence, despite the fact that Catholics have improved their position in the public sector, the unemployment differential has remained.

## THE US COMPARISON

The fair employment legislation, originally conceived in 1973, was enacted and came into force towards the end of 1976. It was, to a considerable extent, modelled on the United States legislation, although by the time it had actually been introduced, substantial changes had taken place in the United States. It is, therefore, useful to compare what has happened in Northern Ireland with what has happened in the United

States. However, one must always bear in mind that the legislation in the United States has been operational for approximately twice as long as the Northern Irish fair employment legislation. It has operated in circumstances where, with the exception of a blip in the 1970s, there has been fairly continuous economic growth and where gradually it has become acceptable to give preferential treatment to underrepresented groups, and, in this aspect, it has been more radical than our legislation.

Smith and Welch (1986: p. ix) have charted what happened to the position of blacks in the United States since the 1950s.

The real story of the last 40 years has been the emergence of the black middle class, whose income gains have been real and substantial. The growth in the size of the black middle class was so spectacular that as a group it outnumbers the black poor. Finally, for the first time in American history a sizable number of black men are economically better off than white middle class America. During the last 20 years alone the odds of a black man penetrating the ranks of the economic elite increased ten-fold.

If one examines the comparative wage rates between black and white one finds the picture shown in Table 8.1. It is a picture of fairly continuous and sustained improvement in the wage rates of blacks versus whites. It is estimated that if one controlled for years of education, the current black wage rate would still leave a deficit of 20%—a substantial figure, but a vast improvement on earlier years.

The changes in the position of blacks in employment are striking. One should not, however, totally forget that they are improvements from a very low base and that there are still vast

TABLE 8.1. *Black wages as a percentage of white wages, United States*

|  | Census year | | | | | 1986 |
|---|---|---|---|---|---|---|
|  | 1940 | 1950 | 1960 | 1970 | 1980 | |
| Percentage | 43.3 | 55.2 | 57.5 | 64.4 | 72.6 | 73.4 |

*Source*: Smith and Welch, 1986, for 1940–80; Bureau of Labour Statistics for 1986.

TABLE 8.2. *Black and white male unemployment rates of those aged 16–64, United States* (%)

| | Census year | | | | 1986 |
|---|---|---|---|---|---|
| | 1950 | 1960 | 1970 | 1980 | |
| Black men | 7.3 | 8.9 | 6.7 | 12.6 | 15.4 |
| White men | 4.0 | 4.5 | 3.6 | 5.8 | 6.8 |
| Black/white ratio | 182.5 | 197.8 | 186.1 | 217.2 | 226.5 |

*Source*: Smith and Welch, 1986, for 1940–80; Bureau of Labour Statistics for 1986.

discrepancies. For example, blacks are almost 10% of the working civil population, but over 40% of servants in private households, over 30% of nursing orderlies, while around 3% of architects, professional engineers, scientists, doctors, dentists, and lawyers. Although blacks are present in overwhelming numbers as nursing orderlies, less than 7% of registered nurses are black.

There is, however, another side of the coin, if one compares male unemployment rates (Table 8.2). This paints a rather different picture. We are met once more with the paradox that the position of blacks in employment has not only not improved, but has actually significantly deteriorated over the period. Indeed, if one compares the Northern Irish figures, one finds that between 1971 and 1985 there was a tiny improvement, but so tiny as not to be significant, in the discrepancies between Catholic and Protestant unemployment, whereas in United States between 1970 and 1986 there was a deterioration of 22%.

It is hardly surprising that William Wilson (1978), in his book *The Declining Significance of Race*, wrote, when comparing the ever worsening situation for unskilled black workers with the increasing opportunities for educated blacks with skills, that 'affirmative action programmes are not designed to deal with the problem of the disproportionate concentration of blacks in the low wage labour market. Their major impact has been in the higher paying jobs of the expanding service producing industries in both the corporate and Government sections'

(Wilson, 1978: 120). He wrote about the deepening economic schism developing in the black community, with the black poor falling further behind middle- and upper-income blacks.

The Committee on the Status of Black Americans, in their book *A Common Destiny* (Jaynes and Williams, 1989), emphasize how the lot of poorer blacks is tied up with economic conditions:

Blacks have a strong interest—stronger than the white majority—in national policies that hold unemployment low and keep the economy expanding vigorously. At the same time, their sensitivity to the nation's macroeconomic performance is a symptom of their continuing marginality and inferiority in economic status . . . The US economy grew rapidly and quite steadily in the first quarter century after World War II. It slowed down after 1973, battered by oil and energy crises, lagging productivity growth, stagflation, government-induced anti-inflationary recessions, high unemployment, high interest rates, financial disturbances, and international competition. Black economic gains were most substantial during the sustained booms of the 1940s and 1960s . . . Since 1970, blacks' relative economic position improved only slowly, and since 1980 it has deteriorated. (Jaynes and Williams, 1989: 294)

Since the new fair employment legislation is, to a considerable extent, based on the United States legislation, with some minor but significant changes, one might be forgiven for saying that we have seen the future and it does not work.

The job of the new Commission, set up under the Fair Employment legislation, will be to make it work and to show that, unlike the situation in the United States, real progress can be made in improving the position of the Catholic unemployed. The new Commission must in particular be aware that it will not be enough to make progress which will result in a further deepening of Wilson's schism between poorer working-class Catholics and those Catholics in middle-class occupations.

There is a number of differences between the Northern Ireland situation and the situation in the United States. The first one, of course, is that our economic position in terms of employment is not nearly so strong, and that whereas we have an unemployment rate today which is two to three times as high as our unemployment rate in 1974, they have an unemployment rate which is comparable to that of 1974. That is on the minus side.

On the plus side there are important differences. The first one is that whereas there are still some educational differences between the two sections of the community in Northern Ireland in terms of qualifications gained by school-leavers, these differences are not enormous by comparison with the gap in qualifications between blacks and whites in the United States (see Chapter 4). The second difference is that we have, unlike the United States, solved the problem of a low level of participation by the minority group in higher-level education. There is no longer a discrepancy between the proportions of Protestants and Catholics who go on to higher education. This is in spite of the fact that Catholics on average come from more disadvantaged economic backgrounds than Protestants. Interestingly, we have solved it not by having any form of preferential treatment for minority candidates for university, which happens in the United States, but by making assistance for students available on a common basis to all who come from a deprived or poorer background. The fact that this problem has been solved is a tribute not only to the system, but also to the Catholic schools who have so successfully overcome the economic disadvantage of their pupils' backgrounds (Cormack, Osborne, and Miller, 1989).

One other major difference is that affirmative action and monitoring arrangements only apply in the United States to those companies with more than 100 employees or those Federal contractors with more than 50 employees. This means that as much as 50% of private-sector employment is not covered by the arrangements. The new fair employment legislation, however, will by 1992 require the monitoring by all employers with more than 10 employees which will amount to 85% of all full-time employees. This difference could be crucial, because it has been written, again by Smith and Welch, that whereas those companies covered by affirmative action in the United States have shown substantial increase in black representation, 'affirmative action resulted in a radical reshuffling of black jobs in the labour force. It shifted black male employment towards Equal Opportunity Commission covered firms and industries and particularly into firms with federal contracts. Reshuffling is the right term, because the mirror image is that black employment in the non-covered sector plummetted' (Smith and Welch, 1986: p. xxi). Because of the much smaller proportion of non-monitoring companies in Northern Ireland that should not be a problem.

Whereas there has been a major effort to enhance access to training in the United States, this is still an area where there are major shortfalls for the black community. In Northern Ireland, with a relatively small economy and with highly sophisticated training arrangements, it should be possible to ensure that there is no shortfall of skills in the Catholic community.

Perhaps the major plus point in Northern Ireland may be that it is accepted by the government, many politicians, the trade union leadership, and many employers, that the discrepancy between Catholic and Protestant unemployment rates is unacceptable and must be eliminated. There is no danger of the syndrome described by Bell that the upward movement of some talented and skilled minority members 'is pointed to by much of society as the final proof that racism is dead—a too hasty pronouncement which dilutes the achievement of those who have moved ahead and denies even society's sympathy to those less fortunate blacks whose opportunities and life fortunes are less promising today than they were 25 years ago' (Bell, 1987: 48). Complacency is dead. It must remain buried.

## THE NEED FOR JOBS

All of this can help to ensure that our legislation can begin to bite in the areas of the deprived and unemployed in Derry, West Belfast, and Newry. One would, however, be guilty of deception if one pretended that the strengthened legislation could in itself bring about an elimination of the major employment discrepancies. The only way in which these discrepancies can be eliminated is by a radical concerted effort on the part of government to ensure that new industry, new investment, and new job opportunities are injected into the areas of high unemployment, which are in the main Catholic areas. The number of jobs needed is not enormous; the number of firms needed to make a significant contribution is not enormous; but government must not believe that alterations to Fair Employment legislation, however welcome, can in themselves make the necessary radical transformation in unemployment discrepancies where the United States legislation has so significantly failed.

It has been argued (Smith, 1988) that targeting high-unemployment areas for industrial regeneration would not in itself lead to a reduction in the difference between Catholic and Protestant unemployment rates. It may be that experience in Britain of attempting, by this strategy, to improve the employment disadvantage of blacks has had limited success. The spatial differences between blacks in Britain and Catholics in Northern Ireland are enormous. Blacks in Britain comprise approximately 5% of the population, concentrated particularly in the inner cities. Industrial regeneration of the inner cities, however, may not in itself improve the position of blacks because the new jobs may mostly be taken up by white people from the suburbs who, because of their greater employment experience, will tend to be at an advantage in the employment market. But factories or projects established in Newry, Derry, Strabane, or West Belfast will inevitably attract majority Catholic labour forces and so reduce the disadvantage experienced by Catholics. For example, three American factories have been established in Derry, and together they employ almost 2,000 people, of whom over 70% are Catholic. Other new projects established in some of the other areas of high Catholic unemployment have compositions which are even more Catholic.

## CONCLUSION

Derrick Bell, in his fantasy entitled *And We Are Not Saved: The Elusive Quest for Racial Justice*, took his title from Jeremiah 8: 20—'The harvest is past, the summer is ended, and we are not saved.' After examining the civil rights strategies which have been used to improve the position of blacks, and after painting a depressing picture of the vast amount which still had to be achieved, he ended up with the conclusion that we must 'seek justice for all through a systematic campaign of attacking

poverty as well as racial discrimination . . . I am now convinced that the goal of a just society for all is morally correct,
strategically necessary and tactically sound' (Bell, 1987:
254–6). A determined attack on the evil of unemployment
accompanied by greatly strengthened fair employment legislation must play a central role in solving the problems of
inequality in Northern Ireland.

# 9

# Law and Employment Discrimination: The Working of the Northern Ireland Fair Employment Agency

*George Applebey and Evelyn Ellis*

The substance of this chapter is based on research carried out between November 1985 and September 1986 for the Standing Advisory Commission on Human Rights (SACHR), which had undertaken a review of 'the adequacy of the coverage and effectiveness of existing laws and institutions in securing freedom from discrimination on the grounds of religious belief or political opinion and furthering equality of opportunity in Northern Ireland' (SACHR, 1987: 2). Central to this review was an examination of the Fair Employment Agency (FEA), which was established under the Fair Employment (Northern Ireland) Act 1976, which enacted many of the recommendations of the van Straubenzee Working Party of 1973 (MHSS, 1973); as such, it represents a unique survivor from the 'power-sharing' attempt of the 1970s. Our research was based in part on a critical and comparative legal analysis of the legislation, and also involved the collation of as much empirical data as we were able to gather in the time available relating to the Agency and the discharge of its functions. In amassing these data, we were greatly assisted by the chairman and staff of the Agency itself, and also by others whose personal or professional experiences had brought them into contact with the Agency, anti-discrimination law, and the various policy issues surrounding equal opportunities in Northern Ireland. This meant we spoke to persons in the relevant government departments, trade unions, lawyers, former Agency members, and politicians. The SACHR additionally received written responses to their review from a large number of individuals and organizations and of these we took careful note.

## DUTIES OF THE FEA UNDER
## THE FAIR EMPLOYMENT ACT

At the outset it is worth looking briefly at the salient features of
the legislation. The van Straubenzee Working Party recommended the setting up of an Agency with wide powers and
responsibilities of investigation, conciliation, and enforcement,
which would be independent of government and would em-
body the most useful features of the machinery created to deal
with discrimination in other jurisdictions (especially Great
Britain, the USA, Canada, and New Zealand). They believed
that there were two basic reasons for needing such an Agency;
first, if the only remedies for breach of the legislation lay in the
courts, individuals would face insuperable difficulties in obtaining sufficient evidence to sustain their claims; and second,
the Working Party did not consider that the first resort in
tackling the problem of discrimination ought to be the courts,
since great emphasis should be placed on persuasion, conciliation, and informality. During the course of our research, we
came increasingly to have doubts about the appropriateness of
too much reliance on conciliation; while it may undoubtedly
sometimes be of value, it may also at other times result in the
ineffective enforcement of the law. Many of our recommendations, therefore, tended in the opposite direction and required a
strengthening of legal powers under the Act and encouragement of greater willingness to have recourse to law where the
need arose.

The FEA's duties were expressed by the legislation as being
'(a) promoting equality of opportunity in Northern Ireland'
and '(b) working for the elimination of discrimination which is
unlawful by virtue of this Act' (Fair Employment (Northern
Ireland) Act 1976, s. 1(1)). It is to be noted that the legislation
did not endow the Agency with the duty of keeping under
review the working of the Act and submitting proposals to the
Secretary of State for amendments where these appeared appropriate to the Agency. This was in contrast to the powers of
the Equal Opportunities Commission (EOC) and Commission
for Racial Equality (CRE) in Great Britain, whose enabling
statutes were drafted some while after the Fair Employment

Act. This omission may, even if only indirectly, have robbed the Agency of a critical edge in the carrying out of its duties and encouraged it in a tacit acceptance that there was no point in complaining about defects which it perceived in the legislation. There was, therefore, a good case for allowing the Agency to make periodic reviews of its legal powers and procedures and to submit its own proposals for reform to government.

## RELATIONSHIP WITH GOVERNMENT

Although the van Straubenzee Working Party stressed the importance of the FEA's independence of government, there were both formal and informal links. The legislation provided, in fact, for the appointment of the FEA's chairman and members by the Department of Economic Development. In addition, the payment of both chairman and members was within the complete discretion of the Department of Economic Development (with the approval of the Department of Finance and Personnel) and the Agency's general funding came from, and was under the scrutiny of, the same two government departments. During the course of our research, in April 1986, the Agency, having for many years claimed with some justification to be underfunded by government, received an increase of 35.8% in its annual grant, bringing it to £364,000. This enabled the Agency to appoint several new members of staff, then totalling fifteen, including both clerical and administrative personnel. By the provision or denial of new funding, the government could thus effectively control the Agency's activities, particularly its law enforcement role.

The Agency was also required to produce an Annual Report on its activities at the end of each financial year and this too involved it in close relations with central government since it had to be made to the head of the Department of Economic Development (Fair Employment (Northern Ireland) Act 1976, s. 1). It is quite clear that there was a close relationship between the Agency and the Northern Ireland government, which is scarcely surprising. Allegations have been made about government involvement in Agency policy, which have of course been denied, over such matters, for example, as the initiation and conduct of investigations. Nevertheless, it is obvious to us that the government did exert a strong but subtle influence on the Agency, in particular through appointments and the provision of finance. It may well be argued that such a position—as, it must be admitted, is the case with many other regulatory agencies—represented an unfortunate compromise, with the government effectively able to appear to distance itself from delicate problems but in reality possessing a significant measure of control over them. Government, in such circumstances, loses much of its accountability but little of its authority.

## LAW ENFORCEMENT

The Agency's most important, or potentially most important, functions lay in the fields of law enforcement and in the promotion of equality of opportunity in the sphere of employment. More specifically, Part III of the Act conferred on the Agency sole jurisdiction to determine 'complaints' of unlawful discrimination contrary to the legislation. Part II enabled it to conduct 'investigations'. The latter might have either of two aims: ascertaining the existence, nature and extent of failures to afford equality of opportunity, or deciding on action to be taken to promote equality of opportunity. According to the Act, investigations could be into the composition, by reference to religious beliefs, of any of five classes of people: (i) employees or applicants for employment; (ii) applicants to employment agencies; (iii) members of or applicants to vocational organizations; (iv) applicants to and the recipients of the services of training bodies; and (v) applicants to and the recipients of qualifications from bodies with power to confer qualifications relevant to employment. Investigations could also be into discriminatory practices affecting such classes of people, including practices discontinued before the time of the investigation so far as relevant for explaining the composition of the class of persons in question at that time (Fair Employment (Northern Ireland) Act 1976, s. 12). The Agency was also obliged to identify and keep under review patterns and trends of employment in Northern Ireland in order to see whether they revealed the existence or absence of equality of opportunity and to help the Agency to decide how best equality of opportunity could be achieved or the reasons for its absence.

Apart from determining complaints and conducting investigations, the Agency's other main function under the legislation was the administration of the 'Declaration of Principle and Intent'. This Declaration, by which the subscriber promised in very bland terms to abide by the principles of the legislation and to co-operate with the FEA, could be made by employers, organizations of employers, organizations of workers, persons engaged in occupations, and vocational organizations. The FEA was required to keep a Register of subscribers which was open to public inspection and each subscriber was entitled to a certificate describing him or it as an Equal Opportunity Organization (Fair Employment (Northern Ireland) Act 1976, ss. 6 and 7). The Agency had three specific powers with respect to the Register; these were to require subscribers to reaffirm their intention to adhere to the Declaration, to remove subscribers from the Register in particular after proceedings against them established a breach by them of an aspect of the legislation, and to restore to the Register a subscriber who had been previously removed.

In order to arrive at a meaningful assessment of both the legislative scheme and the FEA itself, we chose to examine the Agency's powers and record in relation to each of its three main functions as outlined above.

## THE AGENCY AND INDIVIDUAL COMPLAINTS

### *Design of the Legislation*

The Agency's brief was to investigate all complaints of discrimination in employment on the ground of religious belief or political opinion. (Certain important exceptions to this principle will be mentioned below.) Unlike British sex and race discrimination legislation, the Act only outlawed direct discrimination; indirect discrimination (by which was meant the imposition of a condition or requirement which could not be justified and had a significant adverse impact for a particular group) was not unlawful under the Fair Employment Act, although it may have constituted failure to accord equality of opportunity and so be remediable by way of an investigation. Even the definition of direct discrimination may not have been entirely clear to members of the Agency who determined individual complaints. Some confusion arose as a result of a statement by Lord Lowry (the previous Lord Chief Justice) in the Northern Ireland Court of Appeal in *Armagh District Council v. FEA* (IRLR, 1984:234) in which he said that 'although malice (while often present), is not essential, deliberate intention to differentiate on the ground of religion, politics, sex, colour or nationality (whatever is aimed at by the legislation) is an indispensable element in the concept of discrimination'. However, this statement should not be taken to mean that a discriminatory motive is essential to proving direct discrimination. As the CRE in Great Britain has pointed out, 'a person discriminates unlawfully where he or she declines to appoint black staff and does so to meet supposed customer dislike of such staff even though he or she does not share that dislike. His or her *grounds* for doing so are racial even though the *motives* are not' (CRE, 1985). The legislation would have been considerably improved by the addition of a specific statement to the effect that direct discrimination did not necessarily involve a religious or political motive.

Unlike the British anti-discrimination legislation, the only redress for an individual alleging unlawful discrimination was to complain in the first instance to the Agency (Fair Employment (Northern Ireland) Act 1976, ss. 24 and 43). The Agency had to investigate such a complaint, provided only that it was made in writing and was not, in the opinion of the Agency, 'frivolous'. Apart from this power not to proceed at all with a frivolous complaint, the Agency appears to have had no power to do other than proceed and, in particular, it could not abandon an investigation which it initiated if it later became clear that there was no case to answer. This system for complaints was discontinued in the British legislation, essentially for two reasons: first, there is a danger of an agency becoming completely bogged down by the number of complaints, with resulting delay and injustice to individuals; and second, an agency's priorities become determined by the essentially random order in which complaints are received, rather than on the basis of a planned strategy, taking into account patterns and levels of discrimination. The newer British model therefore gives an individual power to take an action directly to the courts or tribunals, but allows the Commissions a strategic role over litigation by giving them power to fund or otherwise to assist complainants. As our research proceeded, we came more and more to the view that these same arguments were equally applicable to individual complaints under the Northern Ireland legislation.

At the completion of the investigation of a complaint, the Agency had to arrive at a finding as to whether or not unlawful discrimination had occurred. At this point, its conciliatory role, as envisaged by the van Straubenzee Working Party, came to the fore. Even if there was no finding of unlawful discrimination, the Agency had to use its best endeavours to secure a settlement of any difference between the complainant and the respondent which was disclosed by the complaint. Where the finding was that unlawful discrimination had been committed, the Agency had to endeavour to secure, if appropriate, a satisfactory written undertaking by the respondent to comply with the terms of the settlement (Fair Employment (Northern Ireland) Act 1976, s. 25). There was a real danger that, in the process of conciliation, the Agency might have been tempted to settle for something less than the successful complainant was entitled to expect and that the appropriateness of conciliation as a remedial tool was exaggerated by this legislation.

If the Agency found unlawful discrimination and a settlement or undertaking had either not been obtained or, not been complied with, then it could serve a notice on both parties containing recommendations as to the action to be taken by the respondent. There were two routes by which the proceedings could then find their way to the county court. First, either complainant or respondent could appeal against the Agency's finding. The Northern Ireland Court of Appeal held in *FEA* v. *Craigavon Borough Council* (IRLR, 1980: 316) that such an appeal was not merely a review of the Agency's finding but a complete rehearing of the complaint. The resulting procedure was not only lengthy and repetitive, but also called into question the whole role of the FEA in the complaints process: if the FEA's findings were so suspect as to merit a complete rehearing of the case, it could be argued that the power to make findings should have been removed from it altogether. On the other hand, if confidence was reposed in the ability of the Agency to reach findings, then a review as to the legality of the procedures adopted by it ought to have been a sufficient guarantee of rights. The second way in which the case could reach the county court was at the suit of the FEA. If the respondent did not comply with his undertaking or with the Agency's recommendations, then the Agency could sue him in tort. Enforcement of this part of the legislation was thus ultimately in the hands of the county court.

### Exceptions from the Act

The legislation contained a number of either non-controversial or else practically insignificant exceptions, for example with respect to employment as a minister of religion, where a particular religion or political opinion was an essential requirement for the job, where the employment was for the purpose of a private household, and in the case of charities and of pre-existing legislation. However, there were also two other exceptions which were both controversial and of great practical significance to the problems of religious division in Northern Ireland.

The first was the well-known exception in respect of employment as a school teacher (Fair Employment (Northern Ireland) Act 1976, s. 37 (1) c). The FEA was required to keep this exception under review, with a view to considering whether it was appropriate that any steps should be taken to further equality in the employment of teachers. In particular, for the purpose of discharging this duty, the Agency was empowered to conduct investigations into the religious composition of the teaching staff of schools and into practices affecting such staff (Fair Employment (Northern Ireland) Act 1976, s. 38). The Agency could report to the Secretary of State on the exercise of its functions in this respect and could make recommendations to him as to any action which the Agency thought should be taken to further equality of opportunity in the employment of teachers. The Secretary of State had power to alter or revoke the exception (Fair Employment (Northern Ireland) Act 1976, s. 39). In practice, the Agency never mounted an investigation under this provision, although it undertook three reviews of the area. It was widely criticized on account of its perceived low level of activity here. However, we believe that there was considerable force in its argument that the exception for school teachers could not be removed from the legislation unless the underlying system of segregated schooling in Northern Ireland was changed; the Agency maintained that removal of the exception would give an unfair advantage to Catholic teachers over Protestants since the Catholic voluntary and voluntary-maintained schools would insist on employing only Catholic teachers, relying on the essential occupational qualification exception, whilst the controlled schools would have no such argument to enable them to reserve posts for Protestants.

The second controversial exception was that relating to national security; the legislation did not apply to an act done for the purpose of safeguarding national security or protecting public safety or public order (Fair Employment (Northern Ireland) Act 1976, s. 42). In addition, it was provided that a certificate signed by the Secretary of State was to be conclusive evidence that an act was done on the ground of national security, public safety, or public order; this exclusion of judicial review (*WLR*, 1986: 1038; Ellis, 1986) meant that there was a very worrying absence of control over the executive in this area. In practice, the government appeared to make fairly extensive use of this exception and, by the date of our research, the FEA had tracked down some seventeen cases in which a certificate had been issued. We consider that the legislation required redrafting in this area, so as to set out far more clearly on what grounds the government was permitted to issue certificates, and to provide for judicial review to ensure that the government did actually act within such constraints.

## The FEA's Record with Respect to Individual Complaints

The number of complaints received by the Agency appeared at first sight to be surprisingly low. By the end of March 1986, by which time it had been in existence for nine and a half years, the FEA had dealt with 527 complaints. In an area with a population of one and a half million, this represents an average of 55.5 complaints per year. Comparison with the rates of complaint to the other anti-discrimination agencies can provide at best only a crude guide, since actual levels of discrimination on different grounds are likely to be different, and also the procedures for the making and recording of complaints vary. However, the Northern Ireland EOC, perhaps the closest comparator to the FEA, had received a total of 913 complaints and enquiries by the end of March 1985 (after nine years in existence). It seems worthy of note that this figure is considerably in excess of the FEA figure, especially given that sex discrimination complaints are not legally required to be channelled through the EOC, although the EOC's remit is wider than employment, with which the FEA is solely concerned.

Again, the statistics relating to the CRE seem significant. The ethnic minority population of Great Britain is about three million so that, at 1985 rates, the CRE annually receives 0.25 complaints per thousand of the population. Taking the Catholic population of Northern Ireland as three-quarters of a million, the FEA handled only 0.07 complaints annually per thousand; if the Protestant population is also added in, the per capita figure of course becomes even lower. All this suggests that either religious and political discrimination was not so prevalent as people often assume or, alternatively, that it was prevalent but that people were reluctant to rely on the machinery provided by the legislation, perhaps because of fear of becoming targets of violence and perhaps also because of a lack of confidence in the efficacy of the law. This latter viewpoint could be justified by looking at complainants' success rates under the legislation.

Of the 527 complaints made to the Agency by the time of our research, only 38 (that is to say, 7.2%) had resulted in findings of discrimination. Taking into account appeals from Agency findings, the total falls to 32 (which represents 10.6% of the cases which at that time had actually proceeded to a finding). Again, the success rate in other anti-discrimination suits provides some sort of yardstick although of course, because of procedural differences, exact comparisons are impossible. The Northern Ireland EOC's figures indicate something in the region of 30 successful sex discrimination claims in employment, but the picture is complicated because many sex discrimination cases are settled before reaching tribunals. Of those in which the EOC grants assistance, the success rate before the tribunals is claimed by the Commission to be about 60%. The average success rate for British sex discrimination claims actually heard by tribunals is 30% and for race claims the figure was 18% in 1984. The success rate in unfair dismissal cases actually heard by tribunals was 28.7% in 1984.

A rate of 10.6%, therefore, appears an extremely meagre yield for claims under the Fair Employment Act. Although by no means providing the only yardstick by which the effectiveness of the legislation can be measured, such a poor success rate must inevitably (if unquantifiably) have operated to discourage would-be complainants from making claims. It also suggests an uneconomic use of Agency resources: complaints were estimated by the Agency to occupy around 35–50% of staff time and this seems a very costly way to have obtained 3.4 findings of discrimination each year.

In our view, there were six major causes for the poor success rate of fair employment claims. First was the absence of any screening process by means of which weak cases could be disposed of at an early stage of the proceedings and without the necessity for a hearing. Second was the methodology of the investigative process. The FEA was placed in a compromised position by the legislation, perceived as an ally of the victim of discrimination at the same time as impartial arbitrator, and this made it very difficult for it to get to the bottom of the facts. Third was the fact that complainants were rarely legally represented, whereas respondents were often so represented. Once again, this resulted from the peculiar position of the FEA under the legislation but, in our opinion, it usually operated to the detriment of complainants. Fourth was the burden of proof which was shouldered by the complainant and which, we believed, would have had to be reversed before complainants stood any serious chance of being able to prove their cases (other than in the most blatant of scenarios). Fifth was the absence of indirect discrimination from the grounds on which a complaint could be made. The concept of indirect discrimination, though in need of refinement (CRE, 1985; EOC, 1986), has proved vital in the fields of sex and race discrimination and we believed it ought to have been added to the Fair Employment Act. And sixth was the process for the adjudication of complaints. The members of the Agency were lay members of the community. They had no necessary familiarity with the law on discrimination, which, as the CRE in particular has argued, is extremely subtle and requires heightened perceptiveness on the part of those who administer it (CRE, 1985).

All these considerations led us to suggest to the SACHR that jurisdiction over fair employment cases should be removed from the Agency and vested in the industrial tribunals, in a similar way to that operating at present in Northern Ireland in sex discrimination claims. We thought the Agency should be given a power akin to that of the EOC's and CRE's to assist individual complaints, either financially or through the giving of advice, which would place it in a strategic position with respect to the development of the law. If the tribunals were able to reach findings in individual cases, they would of course also have to be given the power to award remedies and this would cut out the protracted stages of conciliation and enforcement in the county court which were mentioned earlier.

We are aware of a body of criticism of the record of the industrial tribunals in Great Britain in discrimination claims (CRE, 1985; Leonard, 1987; EOC, 1986). The problem in Great Britain, however, seems to stem from the fact that there are relatively few discrimination cases and a relatively large number of industrial tribunals, with the result that each tribunal chair acquires very little experience of discrimination claims. It is our view that such a situation would not result in Northern Ireland. In part this is because there are more sex discrimination claims per capita in Northern Ireland than in Great Britain; but more importantly, because of the small size of Northern Ireland, it is possible in practice for tribunal chairs to develop their own expertise. Indeed, at present, all sex discrimination claims are dealt with by tribunals chaired by one out of a small band of full-time chairs. There is no reason why such a practice should not be encouraged with respect to fair employment claims so that, in reality, Northern Ireland would possess the specialized tribunal demanded in Great Britain by the CRE.

## THE AGENCY AND PART II INVESTIGATIONS

### *Design of the Legislation*

The legislation was noticeably far less complex in terms of the procedure for investigations than is the sex discrimination and race relations legislation in Great Britain (Applebey and Ellis, 1984). In particular, no distinction was drawn between general investigations and those triggered by a belief that discrimination was occurring; the resultant legal difficulties and opportunities for a respondent to seek judicial review and to hold up the proceedings were therefore reduced. This was greatly to the advantage of all concerned in Northern Ireland, for the British situation, which has become a legal quagmire, is to be neither copied nor envied.

Where an investigation revealed that someone had failed to afford equality of opportunity, the Agency was required to use its 'best endeavours' to ensure that he acts to promote such equality and, if appropriate, to obtain a written undertaking to that effect. If an undertaking was not forthcoming, the Agency could serve a notice on the person in question containing directions for promoting equality of opportunity. This was in line with the van Straubenzee Working Party's view that positive action programmes, both voluntary and required by the FEA, were essential in the Northern Ireland context. If an undertaking was given but not complied with, the Agency could serve directions on the person or apply to the county court for enforcement of the undertaking; the directions themselves could only be enforced by the Agency on an application to the county court. Although enforcement was in the county court, an appeal lay to the Appeals Board against the content of the Agency's directions, which resulted in a confusing multiplicity of fora. We recommended to the SACHR that the legislation be substantially reformed in this respect; if the industrial tribunals were given jurisdiction to determine individual complaints, it would make sense for them also to be given a jurisdiction in Part II investigations. The procedures should be streamlined and jurisdiction to determine appeals in investigations, together with jurisdiction order enforcement in such cases, should in future be vested in the industrial tribunals.

The post-investigation procedure was somewhat different in Northern Ireland from that in Great Britain under the sex and race legislation. The coercive powers of the EOC and CRE are triggered only by a finding by them of a distinct breach of the law; they have no such powers (other than the broad power to mount a formal investigation) in the absence of such a breach, where there is nevertheless inequality of opportunity. Accordingly, their powers are sharper in nature, enabling them to require, by a so-called 'non-discrimination notice', an immediate cessation of breaches of the law, but not allowing them authority to direct how to proceed in the future nor giving them any authority to reach settlements with respondents. It is certainly arguable that this 'policeman-like' role has not been to their long-term advantage, since it may encourage an adversarial attitude on the part of respondents.

### The FEA's Record on Investigations

The FEA started its first investigation using powers under Section 12 of the Act in 1977. This was into the engineering industry in the Belfast area, in the course of which nine employers were investigated. Then in the late 1970s six others, Bass Ireland, Belfast Telegraph, Cantrell and Cochrane, Cookstown and Londonderry (Derry) Councils, and the Northern Ireland Electricity Service, were begun. At the time of our research, the FEA had embarked on twenty-six investigations and four more were in the pipeline. (The question of whether an exercise was technically an 'investigation' was not always clear, and this made exact calculation difficult.) In the course of these investigations, fifty-six companies, organizations, or authorities were looked at and this list included most of the major employers in Northern Ireland.

If one compares this record with other UK agencies, the EOC (GB) had, over a similar period, undertaken nine investigations, the CRE by the end of 1985 had started nearly fifty, and the EOC (NI) had carried out two. Given the staff available to the FEA, twenty-six appeared at first sight a reasonable total. However, many of the Agency's investigations, on reading the published reports, proved minor, indeed insubstantial, when compared with a similar exercise by one of the British agencies. When we first visited the Agency, we found only one officer working on investigations on a full-time basis, with assistance from two other members of staff, although in the past more people have been engaged in this work; we were told that, after a significant increase in the Agency's budget in 1986, new staff were appointed to undertake investigations.

*Strategy for Investigations*

How did the Agency decide whom to investigate? In interviews we were told that the Agency received suggestions from staff themselves, Agency members, and from 'members of the community'. Many investigations were begun as a result of an individual complaint, even an unsuccessful one, to the Agency. Suggestions were first considered by a sub-committee of the Agency, but the actual decision to investigate was taken by the full Agency, which also allocated its priorities and set out future strategies for investigations some time in advance. It is obvious that the strategy was to concentrate on major employers in both the private and public sectors. Unlike the CRE, the FEA did not choose to investigate small employers in the interests of discovering and publicizing bad practices. We were told that the Agency looked to see who was presently recruiting employees as a major factor in undertaking investigations. Areas of the economy in decline did not make a fruitful target for affirmative action programmes. Otherwise no apparent strategy seemed to us to have emerged. The FEA believed that investigations were worth doing (unlike the EOC (NI), which has a different set of priorities) and that they were the best way forward in trying to achieve greater equality of opportunity. Whatever else may be said, the Agency did not shirk from investigating the most obvious targets, including the Northern Ireland Civil Service. However, it may not have put enough of its personnel into this effort, most of its resources being devoted to complaints.

*Time Spent on Investigations*

The period it took to complete investigations varied enormously: from one to eight years depending on the complexity of the issues involved and the resources available for the investigation. Quite often the report appeared some time after the work of investigating had been completed. The largest of the FEA's investigations, that into the Northern Ireland Civil Service, clearly came into this category. The shortest time was for the investigation into Unipork, where in fact the FEA had technically been invited to do an investigation. In fairness, it seems inevitable that investigations were not something which could be done quickly.

*The Practice of Undertaking Investigations*

The decision in principle to investigate could have been taken some time previously but once it was decided actually to launch an investigation a notice of investigation was sent out. This set out the reasons for doing the investigation, the scope and purpose of the investigation, and normally the relevant sections of the Fair Employment Act. The investigating officer then set about seeking information about the work-force, which could be sent by the respondent, or otherwise investigating officers would go into the company and collect it themselves from the company's records. The company was not allowed to prevent information concerning an individual's religious background from being given to the FEA, though individuals themselves could not be compelled to provide such information.

The investigating officer of the FEA then examined personnel files, looking at attributes such as age, sex, school attended, address, initial grade and present grade within the company, and made an assessment of religious background. Even without an express statement of the employee's religion, staff were confident of being able to gauge this element from the other information supplied to a high degree of accuracy. Those familiar with Northern Ireland will not be at all surprised by this. Estimates given to us of the accuracy of gauging religion from other information ranged from 95% to 99%. Establishing religious background was, therefore, not seen as a problem by staff.

The officer then looked at the religious composition of the work-force to see if there was any 'disparity' in the numbers of one side or the other. If this proved to be so then the next matter was to seek the reasons why. Did applicants to the employer represent the proportions expected? The officer then looked at the composition of recent groups of applicants. Thereafter, if there seemed to be an imbalance in the work-force, the question was 'how best is equality of opportunity to be achieved?'

The officer most directly concerned with investigations to whom we spoke was surprisingly confident of their ability to assess the religious breakdown of the area surrounding the company being investigated, in spite of the fact that the sectarian map of Northern Ireland looks to the uninitiated like a patchwork quilt. The key was said to be to understand the labour market, which could be roughly ascertained from the distance people travelled to work in each area; this would then enable investigators to estimate the proportion of both religions in the vicinity and also to compare this with the sectarian composition of the work-force being investigated. As well as this, published criteria, such as the census and other surveys, and a knowledge of local geography and residence patterns could all be helpful in building up a picture. The information on sectarian background could usually be gained from files which employers handed over to the Agency. The attitude of most employers was characterized as one of 'grudging acceptance'. In assessing the composition of the work-force, staff claimed they did not adopt quotas, nor did they support their introduction. The Agency encountered surprisingly few challenges to its powers and procedures on investigations.

Staff felt that investigations were the way to get change, looking at the whole organization rather than one individual complaint at a time.

Once the investigation was complete, the staff involved then had to organize their conclusions about the composition of the work-force, and in particular about whether there had been an imbalance in recruitment or promotion within the organization. The first draft of the report was then prepared by the staff member in charge. Thereafter, a draft of the report went to the Chairman and the Research Sub-Committee for discussion. A second draft was prepared and the recommendations and findings sent to the employer. The employer was then given time to reply and challenge any findings of fact. Not surprisingly, most did. The report which finally appeared was therefore a result of some interchange between the FEA and the company investigated. It was said that the draft which was sent could, but did not always, contain conclusions.

There was a lot of criticism that reports of FEA investigations appeared after such a dialogue to be the result of negotiation and were, in fact, a version agreed between the FEA and the employer, in which the facts set out may not necessarily have been those found during the original investigation at all. A former member of staff of the Agency claimed that the report on the Civil Service investigation took two years to prepare and went through eighteen drafts. In our view, there is nothing intrinsically wrong with the Agency showing a draft of the report to those investigated. However, a report which has been 'doctored' as a result of negotiations between the Agency and the respondent would lend itself to the charge of dishonesty.

### The Agency's Conclusions

Once the investigation was completed, if the Agency was of the opinion that the person concerned failed to afford equality of opportunity, either generally or in relation to any class of person, it had to use its 'best endeavours' to ensure that he took such action for promoting equality of opportunity as was 'reasonable and appropriate'. This could include, where appropriate, a written undertaking that such action would be taken. If the undertaking was not given or complied with, then the Agency could serve a notice with directions 'for the abandonment, or for the modification in accordance with any instructions given in the directions, of any practice which results or may result in failure to afford equality of opportunity or for the substitution or adoption of new practices specified by the Agency' (HMSO, 1976: section 13). Alternatively, if an employer gave an undertaking but failed to comply with it, the FEA could take the employer to the county court under Section 15 of the Act for the enforcement of the undertakings and directions.

At first sight this was one of the vaguest areas of the FEA's activities. Merely reading the reports makes it difficult to know whether or not it found failure to afford equality of opportunity, whether it secured written undertakings from employers, and what the status of its 'recommendations' really was. There was, of course, nothing in the Act to require the Agency to make a finding or even to issue reports.

Employers were indeed allowed to see the Agency's recommendations in draft form, and these recommendations would go to the Chairman and Research Sub-Committee, though 'discussion' rather than 'negotiation' was the word used to describe the process of arriving at the final wording. Staff involved seemed to feel they had considerable discretion in framing recommendations 'as widely or as narrowly as required': whatever was needed 'to match the problem'.

The Agency's failure to state clearly its findings in its reports and to make clear the legal status of its recommendations were two of the recurring and most emphatic criticisms of the Agency in the submission from the Haldane Society, the most lengthy and cogently argued of the written submissions to the SACHR review. For our own part, the true nature of many of the reports was certainly hard to establish: was this a Section 12 investigation, or a research exercise under Section 11 'reviewing patterns and trends in employment'?

The answer of the Agency to these questions was one of the most interesting results of our research. The Agency pointed out that the Act did not require them to make a 'finding' at the conclusion of a Section 12 investigation (unlike under

Section 25) and, believing that their primary duty under the legislation was to use their 'best endeavours' to promote equality of opportunity, they considered that this could best be achieved by stressing that the results of the investigation were merely the Agency's opinion. Throughout its history, therefore, the FEA never made what would in another context amount to a technical legal finding of failure to afford equality of opportunity, after an investigation, although of course it was frequently of the opinion that the facts revealed inequality of opportunity. The Co-ordinating Director of the FEA gave us this verbal explanation during the course of our research:

The difference between the Agency's forming an opinion (its 'finding') as required by Section 25 and being of the opinion that equal opportunity is not being afforded as outlined in Section 13 is crucial. This is not to say that the opinion of the Agency should be less well founded or that it cannot be challenged, but it is vitally important that the standard of proof in a legal sense should not be set high. It should only be necessary to show that the Agency is justified in its opinion, taking into account:

1. the expertise of the members of the Agency and of its staff,
2. the information obtained,
3. reasonableness of the inferences drawn,
4. the attitudes of the parties involved.

In a finding of unlawful discrimination a specific action or series of actions have to be established as having caused less favourable treatment of an individual.

In being of the opinion that equality of opportunity has not been afforded, the Agency has only to believe in good faith that a person of any religious belief has not the same opportunity as a person of like suitability but of another religion.

It is much easier to show justifiable reasons for being of this opinion than it is to show that specific actions were taken for an unlawful discriminatory reason. It is much more difficult to show that this opinion is not justified, when measures can be taken to make the opportunities of the persons in question more equal.

In addition it will be noted that the opinion of the Agency is advised to the party being investigated by means of a preliminary report, and a rebuttal is invited which the Agency takes into account before finally making its recommendations.

Finally, the question of natural justice plays a much lesser part in matters of equality of opportunity. The Act specifically states that unintentional failure to afford equality of opportunity is included in the term (Section 57 part 4). It is not necessarily blameworthy. Therefore, a party facing directions must show it not only does not knowingly fail to afford equality of opportunity, but that there are no precautions it needs to take to minimize the risk of unintentional failure.

Showing a copy of its preliminary draft was therefore a tactic in using 'best endeavours' to get agreement from employers. According to the Agency the justification for this approach was that it worked. All employers except one accepted or agreed to some recommendations designed to improve equality of opportunity. The Agency always sought a written undertaking, and its recommendations, it was claimed, frequently used goals and timetables. This approach also dramatically lessened the chances of an appeal.

In summary, therefore, the Agency tactic was to use their 'best endeavours' in a fairly non-legalistic way to get results. They succeeded in getting recommendations accepted by almost all the respondents investigated largely by avoiding using their stronger legal powers. This was the *quid pro quo* for not being challenged all the way through the courts, as has happened to the CRE. In our view it was probably not as unreasonable a strategy as it seems to many, especially amid the 'troubles' of Northern Ireland, but only if it worked. The acid test of this approach was whether monitoring revealed that greater equality of opportunity really was being achieved and this, sadly, was in most cases difficult for us to ascertain, although of course many argue strongly that it was not being achieved. The truth is only revealed by results, and in a depressed economy it is difficult to be clear about this. Therefore the question still remains to some extent open as to whether or not the Agency in fact achieved results by this means.

## THE 'DECLARATION OF PRINCIPLE AND INTENT'

The number of subscribers to the Declaration remained under 3,000 until the government made clear (in January 1981) that government contracts would in future normally be awarded only to subscribers. The FEA had been pressing for this for some time; indeed, the van Straubenzee Working Party had, as far back as 1973, expressed strong support for a system of contract compliance. We heard criticism of the Agency over the deal it eventually struck with the government over contracts: the Agency argued that would-be government contractors ought not only to have to sign the Declaration, but also to provide the Agency with information relating to both its past and present employment practices, so that the Agency would be enabled to monitor the organization in question. The government was not keen to include the latter provisions, and critics argued that the Agency was insufficiently aggressive over the issue. After the government announcement, the number of subscribers rose sharply and included most public-sector bodies (although only a minority of District Councils).

However, mere signature of the Declaration was not an end in itself. It was only worth something from the point of view of the aims of the legislation if subscribers felt themselves constrained actively to seek to achieve equality of opportunity within their own organizations. It follows that two in particular of the FEA's powers were especially significant: the power to monitor the actions of subscribers and the power to strike off the Register those subscribers who did not comply with the legislation. In practice, the second of these powers was hardly used by the Agency.

As far as monitoring subscribers was concerned, the Agency's policy was merely to select major companies which signed the Declaration some years ago and write to them asking for details of their policies and the composition of their labour force. Although most of the companies concerned replied, none gave details of labour force composition. It was the Agency's view that this sort of information was more likely to be forthcoming if it conducted oral, rather than written, negotiations. Nevertheless, it believed that it would continue to encounter intense resentment on the part of employers at the notion of categorizing their employees on the basis of religion. It appeared to us that compositional analysis of the work-force was only likely to become a regular occurrence in Northern Ireland where some form of inducement was offered to encourage employers to embark on it or through legal obligation.

The conclusion has to be drawn, therefore, that the Agency's efforts to monitor the activities of subscribers to the Declaration were ineffectual and consequently half-hearted. Since the Agency was in effect the only body with power to undertake this function, this meant that signing the Declaration became a mere paper exercise and carried no assurance whatever that the subscriber really respected the principles of the legislation.

We concluded that there were two principal reasons for the Agency's poor performance in this field. The first was lack of finance, since the effective review of several thousand subscribers was obviously a major undertaking. The situation would have been eased, of course, if the Agency were to have ceased to determine individual complaints. But even more important than lack of finance was the vagueness of the criteria to which the Agency was required to pay attention. It was well-nigh impossible for the Agency to carry out any sensible monitoring in the complete absence of any parameters by which equality of opportunity could be judged. The Agency itself could perhaps be criticized for not being sufficiently aggressive in achieving a definition of equality of opportunity but a much more constructive approach would have been to redraft the Declaration so as to incorporate into it considerably more specific undertakings on the part of subscribers. The Agency's job of monitoring could then have become a live possibility, because it would have known precisely what it had to examine.

A sizeable portion of our report to the SACHR therefore dealt with the possible terms of such a new Declaration. We suggested, after consideration in particular of the US Federal Contract Compliance Program, the GLC/ILEA Scheme, and the MacBride Principles, that each subscriber ought in future to have to give the following specific undertakings:

1. to abide by the letter and spirit of the Fair Employment Act; in addition, where the organization consists of 25 or more persons;

2. to carry out an analysis of its work-force at intervals of two or three years to ascertain its religious composition;
3. at the same intervals, to analyse the major job classifications within the organization and compare the numbers of Protestants and Catholics therein with the local labour force;
4. if these figures compared unfavourably, to draw up a positive action plan to correct the deficiencies, specifying goals and timetables;
5. to send an account of the data acquired through these regular analyses to the FEA for it to examine;
6. to appoint an executive to supervise the positive action programme;
7. to declare commitment to equal employment opportunity in all internal and external policy statements.

Further, we recommended that neither government contracts nor any form of public funding should be receivable by any organization not subscribing to the Declaration. The FEA should have the task of monitoring the organizations on the Register, but monitoring would then be a practical possibility because the Agency would be examining clearly defined issues and would possess the necessary information by means of which to reach conclusions.

## CONCLUSIONS

Since our report was presented to the SACHR, significant developments have taken place in this field. First was the publication in September 1986 of a Department of Economic Development Consultative Paper entitled *Equality of Opportunity in Employment in Northern Ireland: Future Strategy Options* (HMSO, 1986). (For comment on this document, see Ellis, 1987.) Whilst disappointing in its lack of detailed analysis of the existing law and powers of the FEA, this was encouraging in its forthright acceptance of the facts of sectarian inequality in the Province and in its commitment—as regards the public sector at least— to the monitoring of work-forces and to the sanctions of contract and grant denial. The Consultative Document was followed up with a new, vigorously worded, and attractively presented *Guide to Effective Practice* (DED, 1987c) published by the Department of Economic Development and aimed at employers (Applebey, 1988).

The SACHR's report on fair employment was itself completed in the autumn of 1987 and laid before Parliament in October of that year. This densely packed document contains thorough analysis both of the empirical evidence collected by the Policy Studies Institute in its various recent surveys of employment in Northern Ireland, and also of the legal issues surrounding the fair employment legislation. The SACHR made a number of wide-ranging and radical recommendations, many of which follow very similar lines to the conclusions of the present authors, in particular as regards indirect discrimination, the hearing of individual complaints by industrial tribunals, and the introduction of a new and much more precisely worded Declaration to be backed up by the sanctions of central government contract and grant denial. In some matters, the SACHR would have gone rather further than us, in particular in recommending the abolition of the existing FEA and its replacement by a new body with revised powers.

The SACHR itself asked the government to respond to its report by means of primary legislation, preceded by a White Paper. The answering White Paper, *Fair Employment in Northern Ireland*, was published in May of 1988 (HMSO, 1988a). A Fair Employment (Northern Ireland) Bill, implementing this White Paper is, at the time of writing (1989), before Parliament. The Bill itself is the subject of detailed analysis by Christopher McCrudden in Chapter 10. In many respects it is to be welcomed, because it accepts arguments and criticisms voiced both by the SACHR and by the present authors. So, for example, the hearing of individual complaints is to be transferred to a Fair Employment Tribunal, under the same presidency as the Industrial Tribunals, and this body will also replace the Appeals Board. However, the important procedural aspects of the new Tribunal are left by the Bill to be determined at a future date by delegated legislation. Individual complaints are thus to be distanced from the Agency, which itself is to be re-created as the Fair Employment Commission and given greater resources. Indirect discrimination is to become unlawful and may be the subject of an individual complaint but, regrettably, the definition adopted by the new legislation is the same as that under the present sex discrimination and British race relations laws which has been seen in practice to be seriously defective. Most importantly, the Bill contains the potential—but it is only the potential—to transform the old provisions concerning the Declaration into a really useful vehicle for the monitoring of work-forces and the launching of affirmative action plans. After an initial transitional period, all employers of 10 or more people must be registered with the Commission if they are to be eligible for public authority contracts or for government grants. However, the Bill leaves to later delegated legislation the task of spelling out precisely what information is to be required from employers, what comparisons they will be required to make, and what standards they must achieve in order to maintain their registration. It scarcely needs to be pointed out that these issues are critical to the effectiveness of the new scheme.

The FEA has become a part of history. That it played an important part in shaping the law on equal opportunities in Northern Ireland cannot be doubted. As the first institution in this highly sensitive field, it is understandable that it went about its business with a certain amount of trepidation. It was also, as has been seen, shackled with a legislative regime which had to a large extent been demonstrated to be unworkable in Britain. Its experience of operating the legislation has certainly enabled lessons to be learnt and relearnt. Whether the new Commission will emerge from its metamorphosis as a highly effective body, and whether the new anti-discrimination laws will prove more successful than their predecessors, can only be judged in the course of the next few years. It would be unfair to brand them a failure if they do not cure all Northern Ireland's equal opportunities problems; no law can do that and it would be unreasonable to measure its potency wholly in terms of the unemployment rate amongst Protestants and Catholics respectively. However, anti-discrimination laws have a vital deterrent, educative, and exemplary role and they set important social standards. The success or failure of the new legislation, seen in these terms, depends to a large part on the government's resolve and on the content of its delegated legislation. It is in the government's own hands whether it now creates machinery for the stern enforcement of the anti-discrimination laws and the real encouragement of equality of opportunity; or whether, conversely, the exercise is a merely cosmetic one, designed to appease vested interests and encourage investment from abroad rather than to protect the rights of the individual.

# 10

## The Evolution of the Fair Employment (Northern Ireland) Act 1989 in Parliament

*Christopher McCrudden*

### THE PROBLEM

Greater equality between Catholics and Protestants in the distribution of economic resources remains a key element in the reconstruction and reform of Northern Ireland. The most authoritative recent study of the extent of inequality of opportunity in Northern Ireland between the two sections of the community was published in 1987 by the Policy Studies Institute (Smith, 1987*a*, 1987*b*; Chambers, 1987). The single most dramatic illustration of the continuing dimensions of the inequality is to be found in comparing the Catholic and Protestant unemployment rates: Catholic males were two and a half times more likely to be unemployed than Protestant males. According to the Policy Studies Institute study, Catholic male unemployment has continued despite there being over 100,000 job changes a year. This has been a consistent finding of other research (Cormack and Osborne, 1983; Osborne and Cormack, 1986; Eversley, 1989; Rolston and Tomlinson, 1988; Rowthorn, 1987, 1981; Rowthorn and Wayne, 1988).

There are various reasons why it is vital to achieve substantial progress towards reducing this differential within a reasonable period of time. Reasons of social justice may be foremost but there are others. Any future constitutional settlement will depend on increasing the degree of mutual respect between the two communities; mutual respect will not develop without greater equality. It is unlikely that external investment, so necessary for the regeneration of a shattered and depressed economy, will be forthcoming in the absence of greater equality. Unless there is a sharing of the benefits and the difficulties, and unless there is the opportunity to regard major employers as being as much 'ours' as 'theirs', there will be no respect, no accommodation, little investment, and no commitment to the state.

## CIVIL RIGHTS AND THE FAIR EMPLOYMENT ACT 1976

The twentieth anniversary of the first civil rights marches in Northern Ireland took place in 1988. Drawing its inspiration from the American civil rights movement, the campaign has focused on the need to eradicate discrimination against Catholics, particularly in employment (Cameron Commission, 1969; Barritt and Carter, 1972; Boehringer, 1971; Buckland, 1979; Farrell, 1980). This movement led to some action by British governments, both Labour and Conservative. The Northern Ireland Constitution Act 1973 prohibited direct discrimination by public-sector bodies. An influential report (MHSS, 1973) considered the question of discrimination in employment and recommended legislation. Following this, the Fair Employment Act 1976 was passed and a Fair Employment Agency was established in 1977.

The Policy Studies Institute study referred to earlier found that the legislation passed in the mid-1970s had little effect on employers' practices. The vast majority of employers interviewed for the study believed that the Act had made little, if any, impact on their practices and procedures. Job discrimination was still thought to be justifiable in certain circumstances by a considerable number of employers. Informal recruitment and appointment procedures contributed to continuing levels of segregation. Too often, investigations by the Fair Employment Agency appear not to have made an impact beyond the individual organization investigated. Very few establishments were formally monitoring the religious composition of the work-force. Indeed, very few establishments were carrying out any type of equal opportunity measure (Chambers, 1987; see Chapter 6).

There was a number of reasons for the ineffectiveness of the

ere was an almost complete lack of interest in the
Labour and Conservative governments between
0s and the mid-1980s, despite the limited effect that
ion was having; the FEA adopted a restrained
to enforcement; negotiation rather than litigation
was the dominant approach even when it delivered relatively
little; employers' organizations, soon after the legislation came
into effect, adopted a belligerent approach to the FEA; the
Social Democratic and Labour Party, the main constitutional
party representing Catholics, and the trade unions, seldom
adverted to the issue (Bew and Paterson, 1985; Darby, 1987;
McCormack, 1988; McCrudden, 1982, 1983; Mullan, 1988;
Osborne, 1980; Rolston, 1983; SACHR, 1987).

## PRESSURES FOR REFORM

Since the mid-1980s, however, inequality of opportunity between Catholics and Protestants in Northern Ireland has again
become a key political issue, largely due to pressure from
outside Northern Ireland. A campaign in the United States was
begun to bring pressure to bear on American corporations,
state legislatures, and municipal governments with invest-
ments in Northern Ireland to adopt a set of anti-discrimination
principles called the 'MacBride Principles' (see Chapter 1).
The MacBride campaign, despite well-orchestrated opposition
from the British and American governments, has proven popular with state and city legislators. By August 1989 over ten
states had already enacted legislation requiring American
companies in which they invest to ensure fair employment
practices in their Northern Ireland subsidiaries, and many
more were considering similar moves (Booth and Bertsch,
1989; Doherty, 1988; Doyle, 1989; Rubenstein, 1987).

This American campaign began to fill, however partially and
inadequately, the political vacuum caused by the failure of
Northern Ireland's political institutions to address the issue
adequately. Significant American pressure on the issue built up
at the federal level too. A number of Bills were introduced into
the US Congress to accomplish similar goals nationally. Yet
another Bill, sponsored by Congressman Kennedy, which

applied United States federal anti-discrimination requirements to defence contractors in Northern Ireland with which
the American government does business, was passed by the
United States Congress in 1988.

Other important pressures emanating from outside Northern Ireland contributed to a change of approach in government. The commitment of British Labour Party front-bench
spokesmen on Northern Ireland from 1986 ensured that there
were in Westminster influential critics of government inaction.
The Irish government formed by Fianna Fáil became involved
in putting pressure on the British government through the
machinery established under the Anglo-Irish Agreement.

## ADMINISTRATIVE REFORM MEASURES

A number of changes not requiring new legislation were introduced by the British government after 1985 but these were
limited and failed to satisfy criticisms emanating from the
United States, the Labour Party, and the Irish government.
A new advisory *Guide to Effective Practice*, replacing one published in 1978, was issued by the local Department of Economic
Development (DED) in September 1987, but this necessarily
reflected the deficiencies of the Act itself (DED, 1987a). The
resources of the Fair Employment Agency were significantly
increased, but the Agency was compromised by its overly
conciliatory approach. In March 1988 a new Fair Employment
Support Scheme for private-sector employers was launched to
provide employers seeking to implement equal opportunity
programmes with free professional advice and training and
limited financial assistance, but the results of this initiative will
take some time to be demonstrated.

## LEGISLATIVE PROPOSALS

Nothing less than new, effective legislation was likely to satisfy
the external critics and so attention centred on the details of a
new anti-discrimination Act. In September 1986 the local DED
published a Consultative Paper proposing new legislation,
which, while offering some hope of a more robust approach,

still fell lamentably short of what was likely to be effective, particularly given its preference for voluntary compliance rather than enforcement (DED, 1986; McCrudden, 1986). It did, however, succeed in concentrating the minds of others. The Standing Advisory Commission on Human Rights (SACHR) (a statutory committee which advises the Secretary of State for Northern Ireland) had begun a review of the effectiveness of the legislation in January 1985. Stimulated by the prospect of new legislation, the Commission published a Report in October 1987 (SACHR, 1987). This provided the most comprehensive and authoritative analysis (based on re-ports commissioned from the Policy Studies Institute) of the problem and a detailed set of proposals for legislation and other government initiatives. The SACHR's recommendations were supported by members drawn from both communities, from the trade union movement, and from the Confederation of British Industry. From its publication, the report formed the bench-mark against which government's proposals were judged.

In March 1988 the government responded with a revised set of outline proposals for legislation (DED, 1988; McCrudden, 1988a), followed in May 1988 with the more detailed White Paper, *Fair Employment in Northern Ireland* (HMSO, 1988a). There were valuable elements in these proposals, in particular the introduction of contract compliance, a monitoring require-ment, the prohibition of indirect discrimination, and the trans-fer of individual complaints to a specialist division of the industrial tribunals. The White Paper thus showed signs of progress. The government appeared, in effect, to have accepted both the SACHR's analysis of the problem, and the broad outlines of its policy recommendations on how to tackle that problem.

However, in its detail, the White Paper fell short of what was necessary to tackle the problem of inequality of opportunity in Northern Ireland employment. Some elements in the proposals were uncertain (e.g. the definition of indirect discrimination and the scope of permitted affirmative action) or were still to be determined (e.g. the scope of contract compliance). On other issues which the Bill would have to address, the White Paper was almost entirely silent (e.g. the scope of remedies) (McCrudden, 1988*b*; Mullan, 1988; O'Callaghan, 1988).

## THE FAIR EMPLOYMENT BILL OUTLINED

The legislation was published in December 1988. The Bill amended the Fair Employment (Northern Ireland) Act 1976 and conformed to the broad outlines of the White Paper in significant respects. The following were and remained central features of the legislation.

The broad powers of formal investigation which the Fair Employment Agency was granted under the 1976 Act were retained. The Agency became the new Fair Employment Commission (FEC). There was, however, no break in the legal existence and continuity of the Agency, whose staff would be absorbed into the new structure. The Commission would have many of the Agency's existing functions and powers. It would continue to undertake investigations, to review patterns and practices in employment, and where necessary to issue directions which would be enforceable on employers. Appeals would be heard by a new body, the Fair Employment Tribunal, which would be a specialized wing of the industrial tribunals. The Commission would be able to accept binding agreements from employers which would be enforceable if not complied with. It would also be able to draw up a Code of Practice for the first time.

A number of new duties were imposed on employers. Many employers were required to register with the new Fair Employment Commission, periodically to review their practices, and to monitor the religious composition of their work-forces. Breach of these duties was a criminal offence, punishable with fines. Where imbalances were evident, employers were expected to engage in affirmative action. Affirmative action would also be enforceable by the Commission. Contract compliance was placed on a statutory footing and both government contracts and government grants could be withdrawn in cases of persistent and recalcitrant failure to obey the Commission or the Fair Employment Tribunal.

Individual complaints of discrimination were transferred

from the jurisdiction of the Fair Employment Agency to the Fair Employment Tribunal. Indirect discrimination was prohibited explicitly; the Bill provided that where a 'condition or requirement' had the effect of disadvantaging a person of a religious or political group this was unlawful unless the person adopting the condition or requirement was able to show that it was 'justifiable'. The Fair Employment Commission was empowered to assist complainants.

### THE MAJOR CRITICISMS LEVELLED AT THE ORIGINAL BILL

Yet, when it was published, the Bill met a barrage of criticism in Northern Ireland and beyond. The Fair Employment Agency was 'disappointed' (FEA, 1989*b*), the Irish Congress of Trade Unions expressed its 'grave concern' (ICTU, 1989) and regarded the Bill as 'seriously defective in a number of key areas'; the Social Democratic and Labour Party saw it as 'inadequate' (House of Commons Library Research Division, 1989); the Equal Opportunities Commission for Northern Ireland was critical (EOC, 1989); the Irish Government restricted itself to supporting only 'the purpose of the legislation' (House of Commons Library Research Division, 1989); American investors were seriously concerned. The Labour Party described it as 'fundamentally and fatally flawed' (McNamara, 1989).

### *Affirmative Action*

The extent to which the Fair Employment Bill enabled employers to implement programmes of affirmative action and the new Fair Employment Commission to require them had long been regarded as one of the acid tests of whether the imbalances between Catholics and Protestants in employment could be reduced within a reasonable time. The Bill's provisions were regarded as endangering existing programmes and retarding the possibility of more extensive affirmative action in the future.

The Bill defined 'affirmative action' as meaning '(a) the

adoption of practices designed to secure fair participation by members of the Protestant, or members of the Roman Catholic community in Northern Ireland, and (b) the modification or abandonment of practices that have or may have the effect of restricting or discouraging such participation . . .' This was not the definition included in the White Paper, which defined it, instead, as action taken to provide 'a more representative distribution of employment in the workforce . . . and full and equal access to employment opportunities' (HMSO, 1988*a*: para. 3.20). In this respect, the Bill's definition was crucially different from the White Paper's in not stipulating that 'fair participation' equals 'a more representative distribution'. The courts, in such a controversial area, it was thought, might well take the view that 'fair participation' does not equal 'a more representative distribution'. The latter, it would be argued, was results-oriented, while the former could well be seeking merely to guarantee a fair procedure. This uncertainty, it was felt, would, at best, breed litigation. At worst, employers would avoid controversy by doing little.

This problem of definition was, however, rendered almost irrelevant due to the main problem with the approach to affirmative action in the Bill. Neither the provisions encouraging affirmative action, nor the provisions empowering its imposition by the Fair Employment Commission, were treated as general exceptions to the duty not to discriminate directly or indirectly. Without such an exception, the duty not to discriminate was regarded as taking precedence, and the ability to take affirmative action was considered to that extent as restricted.

The Fair Employment Agency had in the past sailed close to the legal wind in advocating and, on occasion, requiring types of affirmative action which might be challengeable under the 1976 Act (SACHR, 1987). It was to be hoped that the legality of such measures would be put beyond doubt by the new legislation. However, instead of resolving the uncertainty in favour of legalizing the types of measures advocated by the Fair Employment Agency, the Bill increased the chances that they would be found to be unlawful. A number of measures recommended by the Agency were called into question. Specially targeting Catholic schools in order to increase the number of Catholics applying for jobs in firms where Catholics are severely under-represented would be unlawful. Special selling measures to minority community groups to increase the awareness of opportunities which are open to them would also be unlawful.

The impact of the failure to provide an exception for such measures was judged to be of considerable importance when set in the context of two further factors. First, individuals would be able to litigate issues before the new Tribunal rather than before the Fair Employment Agency. The fudge which the Agency was able to practise on the affirmative action issue would not, therefore, be possible under the new prohibition of indirect discrimination, in addition to that of direct discrimination, meant that the reach of the prohibition of discrimination had been widened considerably.

The new prohibition on indirect discrimination undermined a number of measures which were clearly lawful under the 1976 Act. Measures which have the effect of increasing the chances of Catholics would be challengeable for the first time. For example, where an employer gave a preference for a job to someone who had been unemployed for a considerable period of time, this was considered to be challengeable as indirect discrimination, because it had the effect of disproportionately disadvantaging Protestants. The employer would have to show that this practice was 'justifiable irrespective of religious belief'.

The increased danger which the prohibition of indirect discrimination posed for affirmative action programmes was recognized by government in the Bill as published, but to a very limited extent. The Bill provided that where a training pro-gramme was adopted which gave preference to the unem-ployed, this would not be indirectly discriminatory. However, where an employer adopted a programme of training the unemployed with the intention also of benefiting the Catholic population, this would be unlawful. So, if the employer adopted this practice in order to secure 'fair participation' by members of the Catholic community, this would be unlawful. Only if he did so without such an intention would it be permitted. The approach taken in the Bill was also out of line with existing legislation in other areas of anti-discrimination law. Race and sex discrimination law enabled employers in Britain to engage in equivalent types of affirmative action which the Bill would make unlawful. Employers were permitted to encourage members of an under-represented racial or gender group to apply for jobs and to provide specially targeted race-specific and gender-specific training. Indeed the previous year, the Sex Discrimination (Northern Ireland) Order 1988 had been amended by the government to make it easier for employers to engage in affirmative action to the benefit of women which the Bill would now make illegal in the context of religion.

### Sex Discrimination Order and Affirmative Action

The drafting of the Fair Employment Bill jeopardized the working of the Sex Discrimination (Northern Ireland) Order 1976 in two respects. First, employers might be dissuaded from operating the positive action provisions of the Sex Discrimination Order if there was a possibility that their actions would be held to be unlawful under the Fair Employment Act. In a number of cases under the Fair Employment Act 1976, the Fair Employment Agency had recommended to employers that they adopt affirmative action for women as part of an affirmative action programme for Catholics. The Opposition argument was that under the Bill such recommendations could not now be made: such affirmative action for women could be rendered unlawful. Second, the terms in which the protection for affirmative action training was framed used potentially indirectly discriminatory criteria for the identification of the class which was to benefit by such training, i.e. unemployment status and length of service. It was argued that this breached the Equal Treatment Directive (EC Council Directive, 76/207/EEC).

### Monitoring

Although there would be a requirement on most employers to monitor the existing composition of their work-force, this requirement would not include recording the religion of employees who worked less than 16 hours per week. The

consequent exclusion of many part-time (mostly women) employees from the coverage of monitoring was seen not only as a
classic example of the invisibility of part-time women workers,
it was also seen as likely to result in monitoring returns of
limited usefulness in terms of devising affirmative action programmes. Second, only employers of over 250 employees would
be required to monitor applications, in addition to current
composition. Third and most important, it subsequently
emerged that employers would be allowed to assess the perceived religious affiliation of their employees in only two ways:
by assessing the schools which their employees attended, or
through self-assessment by the employee himself. When this
became clear, the Fair Employment Agency demanded that the
Bill should state explicitly that employers are permitted to use a
range of information in order to assess perceived religious
affiliation and not be restricted to the two methods (schooling
and/or questionnaire) which the illustrative draft Code of
Practice stated would be stipulated in regulations.

### Individual Remedies

The remedies which were to be available to individuals after
succeeding in establishing discrimination reduced a successful
complainant's right under the 1976 Act in two major respects.
First, the level of damages available was reduced. The new
maximum would be £8,500. This fell well short of what the Fair
Employment Agency had been awarding to some individuals
under the existing Act. Second, the provisions requiring an
employer to engage, re-engage, or reinstate a victim of discrimination would be repealed.

### Contract Compliance

The system of contract compliance included in the Bill was a
significant retreat from that proposed in the government's
earlier Consultative Document and from the SACHR's proposals. The government had originally proposed, and the
Commission had welcomed, the adoption of a form of contract
compliance similar to that adopted in the United States. Under

this system, contract compliance has the function of imposing on a contractor a number of 'best practice' requirements which are additional to those imposed on the mass of employers directly under the rest of the legislation. The Bill, however, adopted a much more limited approach to contract compliance. Under the Bill contract compliance had the rather more restricted function of serving as a last resort sanction to be used on those who would not comply with the minimum standards which the legislation imposes on the mass of employers.

## Indirect Discrimination

The prohibition against discrimination in the existing Act was, as we have seen, extended so as to cover indirect discrimination. However, the definition of indirect discrimination in the Bill was taken almost unamended from that in the Sex Discrimination and Race Relations Acts. This restricted significantly the ability of individuals to challenge institutionally discriminatory practices. The term 'requirement or condition' has been interpreted by British courts as failing to embrace the whole range of discriminatory criteria upon which practices may be based. In some cases there may be no obvious requirement but simply practices, or combinations of factors, which operate with discriminatory impact. British courts have decided that a 'requirement or condition' does not exist unless the practice amounts to a 'complete bar if not met' (IRLR, 1982; IRLR, 1988). An employer would, it was argued (Whitmore, 1989) be permitted, without any possibility of an individual being able to gain redress, to give a preference to someone not from West Belfast.

## National Security

The exception in the 1976 Act for actions done for the purpose of safeguarding national security or protecting public safety or public order was retained. This provided that the Secretary of State had power to issue a certificate which was then conclusive evidence that an act was done for that purpose and prevented an investigation into a complaint in respect of which such a certificate had been issued. It was unacceptable that judicial scrutiny of the Secretary of State's decision to issue a certificate under Section 42 of the 1976 Act justifying refusal of employment to an individual or company on the grounds of national security was not available. Not only did the government thereby risk a breach of Article 6 of the European Convention on Human Rights, it would enact one of the more ridiculous anomalies on the statute book. No independent scrutiny was available in cases in which the applicant alleged religious discrimination, but full judicial scrutiny was available where the case involved claims of sex discrimination.

## PATTERN AND PRACTICE INVESTIGATIONS

The creation of a new Fair Employment Commission with increased powers to investigate would be undermined if the Commission could be entangled in legal red tape when it embarked on strategic enforcement. The Bill did little to protect the Fair Employment Commission from this likelihood. Appeals against Directions issued by the Commission in pattern and practice investigations would be heard by the Fair Employment Tribunal. Appeals on points of law from the Tribunal would be heard by the Court of Appeal. The multistage procedure outlined in the Bill for the enforcement of the Commission investigations and appeals against their results, when combined with the inherent possibility of judicial review, seemed destined to give employers multiple opportunities to use the courts to harass the new Commission.

## GOALS AND TIMETABLES

Despite the commendation for the concept of goals and timetables in the White Paper (HMSO, 1988a, and, earlier, DED, 1987a) it was notable by its absence in the Bill. This was seen as an important omission legally because there was uncertainty as to whether the courts and tribunals would look favourably on goals and timetables adopted without explicit statutory backing. More importantly, however, it was seen as a crucial omission in terms of its political impact. Without explicit mention of goals and timetables the vital issue of testable progress was undermined.

## *Excessive Delegation to the DED*

The DED, which had so singularly failed to achieve a reduction in inequality since its establishment, remained in the driving seat on many of the most important issues in the future, despite the SACHR's recommendation that the Department's role should be taken over by the Northern Ireland Office. The Department was given responsibility for drawing up no less than seven sets of subordinate regulations to give effect to crucial areas of the primary legislation. The areas of monitoring, contract compliance, the code of practice, and the procedure before the Fair Employment Tribunal were all subject to detailed regulation by the Department, after the legislation was passed. But these regulations would not be subject to any effective parliamentary scrutiny. In addition, the Department would be empowered to require any information held by the Fair Employment Commission to be turned over to it on demand. The Department would also appoint the members of the Commission and the Tribunal.

## COMMONS CONSIDERATION OF THE BILL

The Fair Employment Bill was published on 15 December 1978 and given its Second Reading by the House of Commons on 31 January 1989. After an acrimonious debate, the Opposition voted against the Bill, as did the Ulster Unionists, while the Social Democratic and Labour Party abstained. The House of Commons Committee stage lasted for 18 sittings between 7 January and the 21 March 1989. Over 200 amendments were considered and a number of amendments were made at the initiative of the government, which also agreed to reconsider a substantial number of other issues. The Bill then returned to the Commons for Report stage, on 25 May, when further amendments were introduced by the government and further commitments were given.

*Commons Amendments*

During the House of Commons stages there were five significant sets of amendments which made their way into the new Act.
First, the government introduced two important affirmative
action amendments: 'lawful' affirmative action would not be in
conflict with any requirement to provide equality of opportunity. Limited protection was given to employers from being
found liable for unlawful direct or indirect discrimination if
they encouraged members of an under-represented religious
group to apply for employment as part of an affirmative action
programme.

Second, the government introduced amendments which
gave the Fair Employment Commission the power to require
employers to set goals and timetables by which to assess
progress made in implementing affirmative action programmes. Employers were explicitly empowered to set goals
and timetables in the context of the periodic review of their
work-forces which they were obliged to undertake. The Commission was empowered to follow up a notice relating to goals
and timetables by means of seeking further undertakings from
an employer and issuing further directions to him, enforceable
before the Fair Employment Tribunal.

Third, an amendment was introduced to raise the limit on
compensation payable to an individual to £30,000.

Fourth, amendments were introduced regarding the requirement to monitor. An amendment was made to empower the
DED to vary the number of hours in the definition of employee
from 16 hours for the purposes of monitoring. More importantly, an amendment was introduced which purported to allow
greater flexibility in assessing the perceived religious affiliation
of employees, though scepticism was expressed by the Opposition as to whether it achieved this objective. The Minister
agreed to look at the amendment again. An amendment was
subsequently introduced in the Lords to empower the Fair
Employment Commission to specify that an employer could
adopt an alternative method of assessing religious affiliation.

Other amendments of less consequence but some importance were also introduced in the Commons. The Fair Employ-

ment Tribunal was specifically required to announce reasons for its decisions. Various financial penalties payable on failure to comply with various monitoring and registration requirements were increased. Employers were required, in conducting their periodic reviews, to examine not only the reasons for the present composition of their work-force, but also the effect of their practices on those being recruited. A questionnaire procedure was introduced to facilitate access to information by a complainant.

## Third Reading Vote

As a result of these changes and government commitments to introduce further amendments in specific areas of concern, both Labour and the Social Democratic and Labour Party voted in favour of the Bill at the end of the Third Reading, while a number of Ulster unionists voted against.

## LORDS CONSIDERATION OF THE BILL

The Second Reading in the House of Lords took place on 28 June. After that events moved quickly. The Lords Committee stage took place on 11 July, Report stage on 17 July, and the Third Reading on 20 July. When the Bill was considered by the Lords, the Opposition not only criticized what it regarded as the government's reneging on a number of commitments to introduce the promised strengthening amendments, it also criticized a set of surprise amendments which it considered had the effect of weakening the Bill still further. Instead of a Bill with all-party support, both Liberal-Democrat and Labour peers launched vociferous attacks on the government's handling of the issue.

## Lords Amendments

Of all the significant Lords amendments, only one met with the approval of the Opposition. An amendment was introduced to empower the Lord Chancellor to appoint and dismiss the

President, Vice-President, and chairmen of the Fair Employment Tribunal. Previously, as we have seen, the Bill had given
this power to the DED. All other amendments were regarded
with disfavour, ranging from irritation to considerable and
deep-seated hostility. We turn, first, to consider the amendments which the government argued honoured commitments
given in the Commons.

First, as we have seen, the Opposition argued that the
definition of indirect discrimination (as a 'condition or requirement' which has an adverse impact on a community) permitted
employers, for example, to prefer employees who are friends
and relations of existing employees, or to prefer workers to be
from East Belfast. Since the definition of equality of opportunity in the Bill did not permit this, the Opposition regarded
this as an anomaly which would serve to weaken the legislation
and increase confusion. The government remained adamantly
opposed to changing the definition of indirect discrimination,
but agreed to introduce an amendment in the Lords to reduce
the anomaly. To achieve this, the Fair Employment Commission was given a discretion to form an opinion that a person
ought to take action for promoting equality of opportunity
as a result of a decision of the Fair Employment Tribunal in
cases taken by an individual, or as a result of evidence given
in such proceedings. The Commission was given a power to
inform the respondent concerned of its opinion that he should
take action to promote equality of opportunity. The person
so informed could give a written and voluntary undertaking
to the Commission to take the appropriate action. The
Commission was also given an additional duty to keep individual cases under review for this purpose. The Opposition
argued, however, that this was inadequate as a method of
dealing with the problems of the flawed definition of indirect
discrimination.

Second, an amendment was introduced which substituted a
new provision for the widely criticized affirmative action training provision. The Opposition argued that the amendment
dealt satisfactorily with neither of the two problems which they
had identified with the previous provision, i.e. (i) that where an
employer adopted a programme of training under the clause,

with the intention of reducing imbalances in part by targeting the under-represented community, the employer would face allegations of direct discrimination from which the clause did not exempt the employer, and (ii) that the clause used criteria for the identification of the class which was to benefit by such training which were indirectly discriminatory against women. Though camouflaged by the terms of the new clause, both these problems remain.

Third, we have seen that the Opposition argued that there were further potential conflicts between the Sex Discrimination (Northern Ireland) Order 1976 and the Bill. An amendment was introduced which protected positive action programmes undertaken by employers under the Sex Discrimination (Northern Ireland) Order 1976 from being challenged as indirectly discriminatory under the Fair Employment Bill. The effect of this amendment is to provide that where an employer carries out positive action in favour of women, such action will not be unlawful as indirect discrimination under the Fair Employment Bill. Therefore, if an employer embarks on affirmative action for women, not intending that this may also have a disproportionate effect in terms of increasing the proportion of, say, Catholics, this will not be unlawful. The Opposition argued, however, that the protection given was inadequate. Under the Bill recommendations to employers that they adopt affirmative action for women as part of an affirmative action programme for Catholics could not be made. The amendment did not address this problem because it provided only that action taken under the Sex Discrimination Order is not to be regarded as *indirectly* discriminatory. It did not provide a defence as regards direct discrimination.

Fourth, the Opposition had argued in the Commons that a simple amendment should be made to the provisions requiring periodic reviews to be carried out by employers. The Opposition proposed that employers should also be required to review redundancy practices to see whether they furthered or retarded equality of opportunity in the same way as the Bill required other recruitment and promotion practices to be reviewed. The government had promised that a suitable amendment would be introduced. When it appeared in the Lords, however, it addressed the point in an entirely different manner, which the Opposition regarded as baffling.

Other amendments were unexpected, as well as unsatis-factory. An amendment was introduced at the Lords Com-mittee stage which altered the requirement in the Bill that employers of over 250 employees should monitor applications, to a requirement that only *advertised* applications need be monitored. After considerable protests from the Opposition, this amendment was withdrawn.

Other unsatisfactory amendments were not withdrawn, however. An amendment was introduced to limit the type of information which the Fair Employment Commission might require an employer to produce to supplement his monitoring returns. This was regarded by the Opposition as weakening the powers of the Commission as originally included in the Bill. A series of amendments was also introduced which increased the number and width of defences available to employers for failure to provide information to the Commission, and in the context of contract compliance default procedures. Lastly, an amendment was introduced to regulate the conduct of cases in which both sex discrimination and religious discrimination is alleged. The Opposition argued that this had the effect of permitting the suspension of sex discrimination complaints to the detriment of complainants and was contrary to the Euro-pean Communities' Equal Treatment Directive.

## COMMONS CONSIDERATION OF LORDS AMENDMENTS

The Bill returned to the Commons for consideration of the Lords amendments on 26 July, amid considerable criticism from the Opposition. Votes were forced on all major amend-ments as an indication of the depth of Opposition disapproval. The Royal Assent was given on 27 July. (The Fair Employment Act 1989, together with the amended Fair Employment Act 1976 and the subsequent Regulations and Orders, may be found in McCrudden, 1990.)

## ASSESSMENT

A balance sheet of the changes made to the Bill indicates that it emerged a better piece of legislation than when it was first published—somewhat less sloppily drafted, less internally inconsistent, and less legally incoherent. Amendments were introduced which met some of the points of concern expressed at the time the Bill was published, in particular in the areas of compensation for individual complainants, monitoring, and goals and timetables. On the other hand, the government failed to remove a number of the elements in the Bill which were widely criticized as limiting its effectiveness and it was further weakened in several other respects. Important flaws, which go to its very heart, remain in the structure of the legislation, in particular concerning the definition and scope of affirmative action, and the location of departmental responsibility for the issue. Whether these flaws will prove fatal remains to be seen.

The government announced in the Commons that there would be periodic reviews of progress with the chairman of the new Fair Employment Commission. A comprehensive review of the working of the Act after five years will subsequently be conducted by the Central Community Relations Unit. The Fair Employment Support Scheme was extended for another year.

Despite these commitments, however, the government has yet to demonstrate that it places the operation of the legislation in the context of the more comprehensive strategy for change proposed by the SACHR. The starting-point of any such strategy must be the recognition that fair employment legislation is a necessary but not sufficient instrument of policy. A body such as the Fair Employment Commission can play only a limited role in bringing about the scale of change which is necessary. While a strong, effective Commission is a necessity, it will always be relatively marginal and thus has to be accompanied by sustained government action on other fronts.

All government policies (including location of industry and government offices, investment decisions, educational developments, and environmental policies) should be subjected to the test of whether their effect furthers or retards equality of opportunity. To do this requires an effective inter-departmental policy unit which has the ear of the Secretary of State. There is considerable doubt whether there is yet sufficient political will to place equality of opportunity at the centre of governmental policy-making in this way. There is a justified fear that the new Fair Employment Commission, and the enactment of the legislation, will be seen by government as relieving it of this task. To do so will mean that the issue identified at the beginning of this article, namely the unacceptable level of inequality between Catholics and Protestants, will remain a major one well into the next century.

# References

Abella Report (1984), Judge Rosalie Silberman Abella, *Equality in Employment*, Royal Commission Report (Ottawa, Supply and Services Department, Canadian Government Publishing Centre).

APPLEBEY, G. (1988), 'Religious Equality of Opportunity in Northern Ireland', *Civil Justice Quarterly*, 7.

——and ELLIS, E. (1984), 'Formal Investigations: The CRE and EOC as Law Enforcement Agencies', *Public Law*, Summer.

ARMSTRONG, D. (1989), 'An Analysis of the Failure to Accept Grammar School Places in West Belfast: 1985–1988' (M.S.Sc. Dissertation in Irish Studies, Queen's University, Belfast).

ASHTON, D. N., and MAGUIRE, J. M. (1980), 'The Function of Academic and Non-Academic Criteria in Employers' Selection Strategies', *British Journal of Guidance and Counselling*, 8: 2.

AUGHEY, A. (1989), *Under Siege: Ulster Unionism and the Anglo-Irish Agreement* (Belfast, Blackstaff Press).

AUNGER, E. (1975), 'Religion and Occupational Class in Northern Ireland', *The Economic and Social Review*, 7: 1.

BARRITT, D. P., and CARTER, C. (1962), *The Northern Ireland Problem: A Study in Group Relations* (London, Oxford University Press).

*Belfast Telegraph* (1988), 'Shorts, How They Have Become One of the Biggest Employers of Catholics in the Province' (18 Aug.)

BELL, D. (1987), *And We Are Not Saved* (New York, Basic Books).

BELL, G. (1976), *The Protestants of Ulster* (London, Pluto Press).

BEW, P., and PATTERSON, H. (1985), *The British State and the Ulster Crisis* (London, Verso).

——GIBBON, P., and PATTERSON, H. (1979), *The State in Northern Ireland* (Manchester, Manchester University Press).

BIRRELL, D., and MURIE, A. (1980), *Policy and Government in Northern Ireland* (Dublin, Gill and Macmillan).

BLOCK, W., and WALKER, M. (1985), *On Employment Equity: A Critique of the Abella Royal Commission Report* (Vancouver, The Fraser Institute).

BOEHRINGER, K. (1971), 'Discrimination: Jobs', *Fortnight*, May.

BOOTH, H., and BERTSCH, K. (1989), *The MacBride Principles and U.S. Companies in Northern Ireland* (Washington, Investor Responsibility Research Center).

BROWN, C. (1985), *Black and White Britain: The Third PSI Study* (Aldershot, Gower).

BRUCE, S. (1986), *God Save Ulster!* (Oxford, Clarendon Press).

BUCKLAND, P. (1979), *The Factory of Grievances: Devolved Government in Northern Ireland, 1921–39* (Dublin, Gill and Macmillan).

Bureau of National Affairs (BNA) (1986), *Affirmative Action Today: A Legal and Political Analysis* (Washington, BNA).

BURTON, F. (1978), *The Politics of Legitimacy* (London, Routledge and Kegan Paul).

Cameron Commission (1969), *Disturbances in Northern Ireland: Report of the Commission Appointed by the Governor of Northern Ireland* (Belfast, HMSO, Cmd. 532).

Campaign for Social Justice (1969), *The Plain Truth* (Dungannon, Campaign for Social Justice).

Canadian Human Rights Commission (CHRC) (1988*a*), *Annual Report 1987* (Ottawa, Canadian Human Rights Commission).

—— (1988*b*), *Operational Procedures for Ensuring Compliance with Employment Equity* (Ottawa, Canadian Human Rights Commission).

CASTLES, S., and KOSACK, G. (1973), *Immigrant Workers and Class Structure in Western Europe* (London, Oxford University Press).

Central Statistics Office (CSO) (1988), *Annual Abstract of Statistics, 1987* (London, HMSO).

CHAMBERS, G. (1987), *Equality and Inequality in Northern Ireland: The Workplace* (London, Policy Studies Institute).

Commission for Racial Equality (CRE) (1985), *Review of the Race Relations Act 1976: Proposals for Change* (London, Commission for Racial Equality).

—— (1990), *Annual Report for 1989* (London, Commission for Racial Equality).

COMPTON, P. A. (ed.) (1981), *The Contemporary Population of Northern Ireland and Population Related Issues* (Belfast, Queen's University).

—— (1986), *Demographic Trends in Northern Ireland*, Report 57 (Belfast, Northern Ireland Economic Council).

—— and COWARD, J. (1989), *Fertility and Family Planning in Northern Ireland* (Aldershot, Avebury).

CORMACK, R. J., and OSBORNE, R. D. (1983), *Religion, Education and Employment: Aspects of Equal Opportunity in Northern Ireland* (Belfast, Appletree Press).

—— and —— (1985), 'Inequality of Misery', *New Society*, 21 Nov.

—— and —— (1989), 'Employment and Discrimination in Northern Ireland' *Policy Studies*, 9: 3.

—— and —— (1990), 'Employment Equity in Canada and Fair Employment in Northern Ireland', *British Journal of Canadian Studies*, 4: 2.

——and ROONEY, E. (n.d.), 'Religion and Employment in Northern Ireland: 1911–1971' (unpublished paper available from the authors).

——GALLAGHER, A., and OSBORNE, R. D. (1990), 'Catholic Schools and the State: The Scottish Example', *The Irish News* (22 May).

——OSBORNE, R. D., and CURRY, C. A. (1989), *Summary Report of the Secondary Analysis of the 1985 Labour Force Survey and Associated Trailer* (Belfast, Policy Research Institute).

————and MILLER, R. L. (1989), 'Student Loans: A Northern Ireland Perspective', *Higher Education Quarterly*, 43: 3.

————and THOMPSON, W. T. (1980), *Into Work? Young School Leavers and the Structure of Opportunity in Belfast*, Research Paper 5 (Belfast, Fair Employment Agency).

——MILLER, R. L., OSBORNE, R. D., and CURRY, C. A. (1989), *Higher Education Demand Survey, Final Report* (Belfast, Policy Research Institute).

COWARD, J. (1980), 'Recent Characteristics of Roman Catholic Fertility in Northern and Southern Ireland', *Population Studies*, 34: 1.

DARBY, J. (1987), 'Religious Discrimination and Differentiation in Northern Ireland: The Case of the Fair Employment Agency', in R. Jenkins and J. Solomos (eds.), *Racism and Equal Opportunity Policies in the 1980s* (Cambridge, Cambridge University Press).

——and MURRAY, D. (1980), *The Vocational Aspirations and Expectations of School Leavers in Londonderry and Strabane*, Research Paper 6 (Belfast, Fair Employment Agency).

DAVIS, K., BERNSTAM, M. S., and RICARDO-CAMPBELL, R. (eds.) (1986), 'Below-Replacement Fertility in Industrial Societies: Causes, Consequences, Policies', Supplement to *Population and Development Review*, 12.

Department of Economic Development (DED) (1986), *Equality of Opportunity in Employment in Northern Ireland: Future Strategy Options— A Consultative Paper* (Belfast, HMSO).

——(1987a), *Religious Equality of Opportunity in Employment: Guide to Effective Practice* (Belfast, DED).

——(1987b), *Equality of Opportunity in Employment in Northern Ireland: Future Strategy Options* (Belfast, HMSO).

——(1987c), *Guide to Effective Practice* (Belfast, DED).

——(1988), *Religious Equality in Employment: New Government Proposals* (Belfast, DED).

——(1989), *Fair Employment in Northern Ireland: Key Details of the Act* (Belfast, DED).

Department of Economic Development (DED) (1990), *Northern Ireland Competing in the 1990s: A Key to Growth* (Belfast, DED).

Department of Manpower Services (DMS) (1978), *Guide to Manpower Policy and Practice* (Belfast, Department of Manpower Services).

DOHERTY, P. (1988), 'MacBride Efforts: US Map Keeps on Getting Darker', *Fortnight*, July/Aug.

DOYLE, L. (1989), 'Ethical Investors Learn to Flex Their Muscles', *Fortnight*, June.

EASTERLIN, R. A. (1980), *Birth and Fortune: The Impact of Numbers on Personal Welfare* (London, Grant McIntyre).

EDWARDS, J. (1987), *Positive Discrimination, Social Justice and Social Policy* (London, Tavistock).

—— (1989), 'Positive Discrimination as a Strategy against Exclusion' *Policy and Politics*, 17: 1.

ELLIS, E. (1986), 'Can Public Safety Provide an Excuse for Sex Discrimination?', *Law Quarterly Review*, 102, Oct.

—— (1987), 'Equality of Opportunity in Employment in Northern Ireland: The Government's Proposals', *Public Law*, Spring.

Equal Opportunities Commission (1986), 'Legislating for Change? Review of the Sex Discrimination Legislation' (Manchester, Equal Opportunities Commission).

Equal Opportunities Commission for Northern Ireland (EOC) (1989), Press Notice, 'EOC for NI Voices Concerns on Fair Employment Bill', 26 Apr.

*Equal Opportunities Review* (1988), 'SACHR Recommends Sweeping Changes in Northern Ireland Religious Discrimination Law', 17.

Equal Opportunities Unit (EOU) (1986), *Equal Opportunities in the Northern Ireland Civil Service: First Report* (Belfast, Department of Finance and Personnel).

—— (1987), *Equal Opportunities in the Northern Ireland Civil Service: Second Report* (Belfast, Department of Finance & Personnel).

—— (1988), *Equal Opportunities in the Northern Ireland Civil Service: Third Report* (Belfast, Department of Finance and Personnel).

ERMISCH, J. (1982), 'The Labour Market: Historical Development and Hypotheses', in D. Eversley and W. Koellmann (eds.), *Population Change and Social Planning* (London, Edward Arnold).

EVERSLEY, D. (1989), *Religion and Employment in Northern Ireland*, (London, Sage). (Additional tables published by the Fair Employment Agency, Belfast.)

—— and HERR, V. (1985), *The Roman Catholic Population of Northern Ireland in 1981: A Revised Estimate* (Belfast, Fair Employment Agency).

——and KOELLMANN, W. (eds.) (1982), *Population Change and Social Planning* (London, Edward Arnold).

Fair Employment Agency (FEA) (1978), *First Annual Report of the Fair Employment Agency* (Belfast, HMSO).

——(1983), *Report of an Investigation by the Fair Employment Agency for Northern Ireland into the Non-Industrial Northern Ireland Civil Service* (Belfast, Fair Employment Agency).

——(1985), *Report of an Investigation into the Northern Ireland Housing Executive* (Belfast, Fair Employment Agency).

——(1989a), *Report of an Investigation into the Queen's University of Belfast under Section 12 of the Fair Employment (NI) Act 1976* (Belfast, Fair Employment Agency).

——(1989b) *Response of the Fair Employment Agency for Northern Ireland to the Fair Employment (Northern Ireland) Bill* (Belfast, Fair Employment Agency).

Fair Employment Commission (FEC) (1990), *Report of an Investigation into the University of Ulster* (Belfast, Fair Employment Commission).

FARRELL, M. (1980), *The Orange State* (London, Pluto).

FALOONA, M., GILLAN, J., and McWHIRTER, L. (1988), *Survey of Northern Ireland Employers* (Belfast, Policy Planning and Research Unit).

FARLEY, R. (1988), 'After the Starting Line: Blacks and Women in an Uphill Race', *Demography*, 25: 4.

FORD, J., KEIL, T., BRYMAN, A., and BEARDSWORTH, N. (1984), 'Internal Labour Market Processes', *Industrial Relations Journal*, 15: 2.

Ford Motor Company (1987), *Ford: Fair Employment Practices in Northern Ireland* (company document).

FULTON, J. (1988), 'Sociology, Religion and "The Troubles" in Northern Ireland', *The Economic and Social Review*, 20: 1.

GAFFIKIN, F., and MORRISSEY, M. (1990), *Northern Ireland: The Thatcher Years* (London, Zed Books).

GALLAGHER, A. (1989), *Education and Religion in Northern Ireland: The Majority Minority Review No 1* (Coleraine, University of Ulster).

GLAZER, N. (1988), 'The Affirmative Action Stalemate' *The Public Interest*, No. 90.

GOULDNER, A. (1970), *The Coming Crisis of Western Sociology* (New York, Basic Books).

GRAY, J., McPHERSON, A. F., and RAFFE, D. (1983), *Reconstructions of Secondary Education: Theory, Myth and Practice Since the War* (London, Routledge and Kegan Paul).

GUDGIN, G., and ROPER, S. (1990), *The Northern Ireland Economy* (Belfast, Northern Ireland Economic Research Centre).

*Hansard* (House of Lords) (1989), 14 Dec., cols. 1434–61.

HARBISON, J. (ed.) (1989), *Growing Up in Northern Ireland* (Belfast, Stranmillis College).

HARRIS, R. (1972), *Prejudice and Tolerance in Ulster: A Study of Neighbours and 'Strangers' in a Border Community* (Manchester, Manchester University Press).

HEANEY, S. (1990), 'Whatever you say say nothing', in *New Selected Poems 1966–1987* (London, Faber and Faber).

HEATH, A., and RIDGE, J. (1983), 'Schools, Examinations and Occupational Attainment' in J. Purvis and M. Hales (eds.), *Achievement and Inequality in Education* (London, Routledge and Kegan Paul).

HEPBURN, A. C. (1982), *Employment and Religion in Belfast 1901–1970* (Belfast, Fair Employment Agency).

—— (1983), 'Employment and Religion in Belfast: 1901–1951', in R. Cormack and R. Osborne (eds.), *Religion, Education and Employment: Aspects of Equal Opportunity in Northern Ireland* (Belfast, Appletree Press).

—— and COLLINS, B. (1981), 'Industrial Society: The Structure of Belfast, 1901', in P. Roebuck (ed.), *Plantation to Partition* (Belfast, Blackstaff Press).

HMSO (1976), *Fair Employment (Northern Ireland) Act 1976* (London, HMSO).

—— (1988*a*), *Fair Employment in Northern Ireland* (London, HMSO, Cm. 380).

—— (1988*b*), *Fair Employment (Northern Ireland) Bill* (London, HMSO).

—— (1989*a*), *Education Reform (Northern Ireland) Order 1989* (Belfast, HMSO).

—— (1989*b*), *Fair Employment (Northern Ireland) Act 1989* (London, HMSO).

HOBSBAWM, E., and RANGER, T. (1983), *The Invention of Tradition* (Cambridge, Cambridge University Press).

House of Commons Library Research Division (1989), 'Fair Employment (Northern Ireland) Bill 1988–1989', Reference Sheet 89/1.

Industrial Relations Law Reports (IRLR) [1980], *FEA* v. *Craigavon Borough Council*, p. 316.

—— [1982], *Perera* v. *Civil Service Commission*, p. 147.

—— [1984], *Armagh District Council* v. *FEA*, p. 234.

—— [1988], *Meer* v. *London Borough of Tower Hamlets* [1988], p. 399.

Investor Responsibility Research Center (IRRC) (1989), 'US Corporate Activity in Northern Ireland', Proxy Issues Report, Social Issues Service, 1990 Analysis B, 26 December 1989 (Washington, Investor Responsibility Research Center).

Irish Congress of Trade Unions (ICTU) (1989), 'Comments on the Fair Employment (Northern Ireland) Bill', Jan.

JAYNES, G., and WILLIAMS, R. (1989), *A Common Destiny: Blacks and American Society* (Washington, National Academy Press).

JENKINS, R. (1988), 'Discrimination and Equality of Opportunity in Employment: Ethnicity and "Race" in the United Kingdom', in D. Gallie (ed.), *Employment in Britain* (Oxford, Blackwell).

JOHNSON, K. W. (ed.) (1985–6), *Black and Minority Health* (Washington, US Government Printing Office).

KENNEDY, R. (1973), *The Irish: Emigration, Marriage and Fertility* (Berkeley, Calif., University of California Press).

LEONARD, N. (1987), 'Judging Inequality' (London, Cobden Trust).

LIVINGSTONE, J. (1987), 'Equality of Opportunity in Education in Northern Ireland', in R. D. Osborne, R. J. Cormack, and R. L. Miller (eds.), *Education and Policy in Northern Ireland* (Belfast, Policy Research Institute).

LOUGHRAN, G. (1987), 'The Rationale of Catholic Education', in R. D. Osborne, R. J. Cormack, and R. L. Miller (eds.), *Education and Policy in Northern Ireland* (Belfast, Policy Research Institute).

LOURIE, G. C. (1984), 'Moral Leadership in the Black Community', in *Black Leadership, Two Lectures in the W. Arthur Lewis Lecture Series* (Princeton, NJ).

LUSTGARTEN, L. (1987), 'Racial Equality and the Limits of the Law', in R. Jenkins and J. Solomos (eds.), *Racism and Equal Opportunity Policies in the 1990s* (Cambridge, Cambridge University Press).

McCORMACK, J. (1988), 'Faceless Men, Civil Rights and After', in M. Farrell (ed.), *Twenty Years On* (Dingle, Brandon Press).

McCRUDDEN, C. (1982), 'Law Enforcement by Regulatory Agency: The Case of Employment Discrimination in Northern Ireland', *Modern Law Review*, 45.

——(1983), 'The Experience of the Legal Enforcement of the Fair Employment (Northern Ireland) Act 1976', in R. J. Cormack and R. D. Osborne (eds.), *Religion, Education and Employment: Aspects of Equal Opportunity in Northern Ireland* (Belfast, Appletree Press).

——(1986), 'Equal Employment in Northern Ireland', *Equal Opportunities Review*, Nov./Dec.

——(1988a), 'Finding the Flaws in North's New Job Proposals', *Irish Times*, 9 March.

McCrudden, C. (1988*b*), 'The Northern Ireland Fair Employment White Paper: A Critical Assessment', *Industrial Law Journal*, 17.

——(ed.) (1990), *Fair Employment Handbook* (London, Industrial Relations Services).

McNamara, K. (1989), 'Flawed Bill Cannot End NI Jobs Bias', *Irish Times*, 31 Jan.

McWhirter, L. (1989), 'Longitudinal Evidence on the Teenage Years', in J. Harbison (ed.), *Growing Up in Northern Ireland* (Belfast, Learning Resources Unit, Stranmillis College).

——Duffy, U., Barry, R., and McGuinness, G. (1987), 'Transition from School to Work: Cohort Evidence on the Youth Training Programme', in R. D. Osborne, R. J. Cormack, and R. L. Miller (eds.), *Education and Policy in Northern Ireland* (Belfast, Policy Research Institute).

——Thompson, D., Duffy, U., and Gillan, J. (forthcoming), 'Gender Differences and the Youth Training Programme', in M. Maguire (ed.), *Unequal Labour: Women at Work in Northern Ireland* (Belfast, Policy Research Institute).

Maguire, M. (1990), *Work, Employment and New Technology: A Case Study of Multi-National Investment in Northern Ireland* (Belfast, Policy Research Institute).

Manwaring, T. (1984), 'The Extended Internal Labour Market', *Cambridge Journal of Economics*, 8: 2.

Marshall, T. H. (1963), 'Citizenship and Social Class', in *Sociology at the Crossroads and Other Essays* (London, Heinemann).

Miliband, R. (1987), 'Class Analysis', in A. Giddens and J. Turner (eds.), *Social Theory Today* (Cambridge, Polity Press).

Miller, R. L. (1983), 'Religion and Occupational Mobility', in R. J. Cormack and R. D. Osborne (eds.), *Religion, Education and Employment: Aspects of Equal Opportunity in Northern Ireland* (Belfast, Appletree Press).

——and Osborne, R. D. (1983), 'Religion and Unemployment: Evidence from a Cohort Survey', in R. J. Cormack and R. D. Osborne (eds.), *Religion, Education and Employment: Aspects of Equal Opportunity in Northern Ireland* (Belfast, Appletree Press).

————Cormack, R. J., and Williamson, A. P. (1990), 'Higher Education and Labour Market Entry: The Differing Experience of Northern Irish Protestants and Catholics', Research Paper 1, Centre for Policy Research (Coleraine, University of Ulster).

Ministry of Health and Social Services (MHSS) (1973), *Report and Recommendations of the Working Party on Discrimination in the Private*

*Sector of Employment* (The van Straubenzee Report) (Belfast, HMSO).

MORRIS, C., COMPTON, P., and LUKE, A. (1985), 'Non-Enumeration in the 1981 Northern Ireland Census of Population', PPRU Occasional Paper No. 9 (Belfast, Policy Planning Research Unit).

MOYNIHAN, D. P. (1986), *Family and Nation* (San Diego, Calif., Harcourt, Brace and Jovanovich).

MULLAN, K. (1988), 'Reforming Anti-Discrimination Legislation: Fair Enough?', *Public Money and Management*, Spring/Summer.

MURRAY, D. (1985), *Worlds Apart: Segregated Schools in Northern Ireland* (Belfast, Appletree Press).

MYRDAL, G. (1944), *An American Dilemma* (New York, Harper and Row).

National Audit Office (1987), *Department of Employment and Manpower Services Commission: Employment Assistance to Disabled Adults* (London, HMSO).

NELSON, S. (1984), *Ulster's Uncertain Defenders* (Belfast, Appletree Press).

*New York Times* (1987), 'From Dissenting Opinions', 26 March.

Northern Ireland Civil Service Commission (1987), *Annual Report 1986* (Belfast, HMSO).

Northern Ireland Economic Council (NIEC) (1981), *Employment Patterns in Northern Ireland, 1950–1980* (Belfast, NIEC).

——(1990), *The Private Sector in the Northern Ireland Economy*, (Belfast, NIEC).

Northern Ireland Economic Research Centre (NIERC) (1989), *Job Generation in Manufacturing Industry, 1973–1986* (Belfast, NIERC).

Northern Ireland Office (NIO) (1985), Parliamentary Question, Northern Ireland Information Service, 3 July.

——(1990) 'Equal Opportunity Proofing of Policy Making', Northern Ireland Information Service, Newsrelease, 9 Mar.

NOVAK, M. (ed.) (1987), *The New Consensus on Family and Welfare* (Washington, American Enterprise Institute for Public Policy Research).

O'CALLAGHAN, D. (1988), 'From Belfast to Brixton: Could New Monitoring Measures Cross the Irish Sea?', *Personnel Management*, Aug.

Office of Population Censuses and Surveys (OPCS) (1980), *Recorded Internal Population Movements; International Migration Monitors*, Series MN (London, HMSO from 1980).

OSBORN, A. E., BUTLER, W. R., and MORRIS, A. C. (1984), *The Social Life of Britain's 5 Year Olds*, (London, Routledge and Kegan Paul).

OSBORNE, R. D. (1980), 'Fair Employment in Northern Ireland', *New Community*, 8.

—— (1981), 'Equality of Opportunity and Discrimination: The Case of Religion in Northern Ireland', *Administration*, 29.

—— (1985), *Religion and Educational Qualifications in Northern Ireland* (Belfast, Fair Employment Agency).

—— (1986), 'Segregated Schools and Examination Results in Northern Ireland: Some Preliminary Research', *Educational Research*, 28: 1.

—— (1987), 'Religion and Employment', in R. Buchanan, and B. Walker (eds.), *Province, City and People* (Belfast, BAAS/Greystone Press).

—— (1989), 'The Funding of Separate (Catholic) Schools in Ontario' (available from the author).

—— (1990), 'Equal Opportunities in the Northern Ireland Civil Service', *Public Money and Management*, 10:2.

—— and CORMACK, R. J. (1986), 'Religion and Unemployment in Northern Ireland', *The Economic and Social Review*, 17:3.

—— and —— (1987), *Religion, Occupations and Employment: 1971–1981*, Research Paper 11 (Belfast, Fair Employment Agency).

—— and —— (1989*a*), 'Fair Employment: Towards Reform in Northern Ireland', *Policy and Politics*, 17: 4.

—— and —— (1989*b*), 'Religion and Gender as Issues in Education, Training and Entry to Work' in J. Harbison (ed.), *Growing Up in Northern Ireland* (Belfast, Stranmillis College).

—— and —— (1990), 'Higher Education and Fair Employment in Northern Ireland', *Higher Education Quarterly*, 44: 4.

—— and MURRAY, R. C. (1978), *Educational Qualifications and Religious Affiliation in Northern Ireland* (Belfast, Fair Employment Agency).

—— GALLAGHER, A. M., and CORMACK, R. J. (1989), 'Review of Aspects of Education in Northern Ireland', in *Fourteenth Report*, Standing Advisory Commission on Human Rights (London, HMSO).

—— CORMACK, R. J., REID, N. G., and WILLIAMSON, A. P. (1983), 'Political Arithmetic, Higher Education and Religion in Northern Ireland', in R. J. Cormack and R. D. Osborne (eds.), *Religion, Education and Employment: Aspects of Equal Opportunity in Northern Ireland* (Belfast, Appletree Press).

—— MILLER, R.L., CORMACK, R. J., and WILLIAMSON, A. P. (1988), 'Trends in Participation in Higher Education in Northern Ireland', *The Economic and Social Review*, 19: 4.

PARKIN, F. (1979), *Marxism and Class Theory* (London, Tavistock).

PAYNE, G., and FORD, G. (1977), 'Religion, Class and Education Policy', *Scottish Educational Studies*, 9: 2.

PEARN, M., KANDOLA, R. S., and MOTTRAM, R. D. (1987), *Selection Tests and Sex Bias*, Equal Opportunities Commission (London, HMSO).

PETERSON, W. (1975), *Population*, 2nd edn. (Toronto, Collier Macmillan).

RAFFE, D. (1984), 'School Attainment and the Labour Market', in D. Raffe (ed.), *Fourteen to Eighteen* (Aberdeen, Aberdeen University Press).

——and WILLMS, D. (1989), 'Schooling the Discouraged Worker: Local Labour Market Effects on Educational Participation' *Sociology*, 23: 4.

ROGERS, R. (ed.) (1986), *Education and Social Class* (Brighton, Falmer).

ROLSTON, B. (1983), 'Reformism and Sectarianism: The State of the Union after Civil Rights', in J. Darby (ed.), *Northern Ireland: The Background to the Conflict* (Belfast, Appletree Press).

——and TOMLINSON, M. (1988), *Unemployment in West Belfast: The Obair Report* (Belfast, Beyond the Pale Publications).

ROSE, R. (1971), *Governing without Consensus* (London, Faber).

ROWTHORN, B. (1981), 'Northern Ireland: An Economy in Crisis', *Cambridge Review of Economics*, 5: 1.

——(1987), 'Northern Ireland: An Economy in Crisis', in P. Teague (ed.), *Beyond the Rhetoric: Politics, the Economy and Social Policy in Northern Ireland* (London, Lawrence and Wishart).

——and WAYNE, N. (1988), *Northern Ireland: The Political Economy of Conflict* (London, Polity Press).

RUBENSTEIN, M. (1986), 'Behind the MacBride Principles', *Equal Opportunities Review*, No. 8.

SEXTON, J. J., and DILLON, M. (1984), 'Recent Changes in Irish Fertility', *Quarterly Economic Commentary* (Dublin, Economic and Social Research Institute).

SIMPSON, J. (1983), 'Economic Development; Cause and Effect in the Northern Irish Conflict', in J. Darby (ed.), *Northern Ireland: The Background to the Conflict* (Belfast, Appletree Press).

SMITH, D. (1977), *Racial Disadvantage in Britain* (Harmondsworth, Penguin Books).

——(1987a), *Equality and Inequality in Northern Ireland Part I: Employment and Unemployment* (London, Policy Studies Institute).

——(1987b), *Equality and Inequality in Northern Ireland: Perceptions and Views* (London, Policy Studies Institute).

SMITH, D. (1988), 'Policy and Research: Employment Discrimination in Northern Ireland', *Policy Studies*, 9: 1.

SMITH, J., and WELCH, F. (1986), *Closing the Gap: Forty Years of Economic Progress for Blacks* (Santa Monica, Calif., Rand Corporation).

Standing Advisory Commission on Human Rights (SACHR) (1987), *Religious and Political Discrimination and Equality of Opportunity in Northern Ireland: Report on Fair Employment* (Belfast, HMSO).

STEVENSON, W. G., MALLON, J. R., and HEPPER, F. (1988), 'Practical Aspects of Monitoring Equality of Opportunity in a Large Organization', *The Statistician*, 37: 3.

TAYLOR, R. (1988), 'The Queen's University of Belfast: The Liberal University in a Divided Society', *Higher Education Review*, 20: 2.

TODD, J. (1987), 'Two Traditions in Unionist Political Culture', *Irish Political Studies*, 2.

TREBLE, J. H. (1979), 'The Development of Roman Catholic Education in Scotland 1878–1978', in D. McRoberts (ed.), *Modern Scottish Catholicism, 1878–1978* (Glasgow, Burns).

WALKER, F. (1986), *Catholic Education and Politics in Ontario*, III (Toronto, Catholic Education Foundation of Ontario).

WARR, P., COOK, J., and WALL, T. (1979), 'Scales for the Measurement of Some Work Attitudes and Aspects of Psychological Well-being', *Journal of Occupational Psychology*, 52.

*Weekly Law Reports* [1986], *Johnston v. RUC* [1986], Vol. 3, p. 1038.

WHITMORE, J. (1989), 'A Helping Hand's Weak Arm', *The Guardian*, 16 June.

WHYTE, J. (1983), 'How Much Discrimination Was There under the Unionist Regime 1921–1968?', in T. Gallagher and J. O'Connell (eds.), *Contemporary Irish Studies*, (Manchester, Manchester University Press).

WHYTE, JEAN, KILPATRICK, R., and McILHENEY, C. (1985), *Are They Being Served? A Study of Young People in the Guaranteed Year of the Northern Ireland Youth Training Programme* (Belfast, The Northern Ireland Council for Educational Research).

WILLMS, D. (1989), 'Pride or Prejudice? Opportunity Structure and the Effects of Catholic Schools in Scotland' (paper presented to the Standing Advisory Commission on Human Rights, Belfast).

WILSON, J. (1987), 'Selection for Secondary Education', in R. D. Osborne, R. J. Cormack, and R. L. Miller (eds.), *Education and Policy in Northern Ireland* (Belfast, Policy Research Institute).

WILSON, T. (1989), *Ulster: Conflict and Consent* (Oxford, Blackwell).

WILSON, W. J. (1978), *The Declining Significance of Race* (Chicago, University of Chicago Press).

WINCH, D. (1978), *Adam Smith's Politics* (Cambridge, Cambridge University Press).

WRIGHT, F. (1973), 'Protestant Ideology and Politics in Ulster', *European Journal of Sociology*, 14.

—— (1987), *Northern Ireland: A Comparative Analysis* (Dublin, Gill and Macmillan).

# Index

Page numbers in bold denote references to tables.

288 *Index*